The Swedish Farmers' Movement and Government Agricultural Policy

Abstract

Micheletti, Michele. Department of Political Science, University of Stockholm, Sweden. *The Swedish Farmers' Movement and Government Agricultural Policy*. New York: Praeger, 1990. ISBN 0-275-93398-9. Monograph. 215 pages.

The book is about the Federation of Swedish Farmers (*Lantbrukarnas Riksförbund*, LRF). Its ability to represent its members and safeguard the interests of farmers in the current period of agricultural reform is the main subject. The historical situation of the LRF and its forerunners, the role played by peasant romanticism in safeguarding the interests of farmers, the evolution of the Agrarian to the Center Party, and the development of Swedish agricultural policy are analyzed. Changes are taking place in agricultural policy. It is expanding to include other issues. Consumers are now a countervailing power in policymaking. There is increased concern over the environment as well as pressure to deregulate. International opinion is important in policymaking. This affects the LRF which is also changing. Its membership has become more heterogeneous. Its goals, strategies, and organizational structures are undergoing transformation. The study analyzes official documents from government, international institutions, and the LRF as well as unpublished sources from the LRF and interviews with policymakers.

The Swedish Farmers' Movement and Government Agricultural Policy

Michele Micheletti

PRAEGER

New York
Westport, Connecticut
London

Library of Congress Cataloging-in-Publication Data

Micheletti, Michele.
The Swedish farmers' movement and government agricultural policy / Michele Micheletti.
p. cm.
Includes bibliographical references.
ISBN 0-275-93398-9 (alk. paper)
1. Agriculture and state—Sweden. 2. Farmers—Sweden—Political activity. I. Title.
HD2018.M53 1990
338.1'8485—dc20 89-16206

Library of Congress Catalog Card Number: 89-16206
ISBN: 0-275-93398-9

First published in 1990

Praeger Publishers, One Madison Avenue, New York, NY 10010
An imprint of Greenwood Publishing Group, Inc.

Printed in the United States of America

The paper used in this book complies with the Permanent Paper Standard issued by the National Information Standards Organization (Z39.48-1984).

10 9 8 7 6 5 4 3 2 1

Contents

Tables and Figures

TABLES

FIGURE

Abbreviations

ARF	Alternativodlarnas Riksförbund (National Association of Biodynamic Farmers)
EC	European Community
EFTA	European Free Trade Association
FNSEA	Federation Nationale des Syndicats d'Exploitants Agricoles
FOA	Försvarets Forskningsanstalt (National Defense Research Institute)
GATT	General Agreement on Tariffs and Trade
JN	Statens Jordbruksnämnd (National Agricultural Marketing Board)
Jo	Jordbruksdepartementet (Ministry of Agriculture)
JoU	Jordbruksutskottet (Standing Committee on Agriculture)
JUI	Jordbrukets Utredningsinstitut (Joint Agricultural Research Institute)
KF	Kooperativa Förbundet (Cooperative Union and Wholesale Society)
KO	Konsumentombudsmannen (Consumer Ombudsman)
KRAV	Kontrollföreningen för Alternativ Odling (National Association for Quality Inspection of Biodynamic Products)
LBS	Lantbruksstyrelsen (National Board of Agriculture)
LO	Landsorganisationen i Sverige (Confederation of Swedish Trade Unions)
LRF	Lantbrukarnas Riksförbund (Federation of Swedish Farmers)
LUI	Lantbrukets Utredningsinstitut (Agricultural Economics Research Institute)
MD	Marknadsdomstolen (Market Court)
NBC	Nordens Bondeorganisationers Centralråd (Central Council of the Nordic Farmers' Associations)
NFU	National Farmers' Union (Great Britain)
NO	Näringsfrihetsombudsmannen (Office of the Competition Ombudsman)
OECD	Organization for Economic Cooperation and Development
RLF	Riksförbundet Landsbygdens Folk (National Farmers' Union)
RRK	Riksdagens Revisorers Kansli (Office of the Parliamentary Auditors)
RRV	Riksrevisionsverket (National Audit Bureau)
SÄ	Svenska Ägghandelsförbundet (Swedish Egg Marketing Association)
SACO	Sveriges Akademikers Centralorganisation (Swedish Confederation of Professional Associations)
SACO/SR	Centralorganisationen SACO/SR (Swedish Confederation of Professional Associations)

SAF	Svenska Arbetsgivareföreningen (Swedish Employers' Association)
SAL	Sveriges Allmänna Lantbrukssällskap (Swedish Agricultural Society)
SFF	Sveriges Föreningsbankers Förbund (Federation of Swedish Cooperative Banks)
SIND	Statens Industriverk (National Industrial Board)
SL	Sveriges Lantbruksförbund (National Federation of Agricultural Cooperatives)
SLIM	Sveriges Livsmedelsindustriförbund (Federation of Swedish Food Industry)
SLR	Svenska Lantmännens Riksförbund (Swedish Farmers' Supply and Crop Marketing Association)
SMR	Svenska Mejeriernas Riksförbund (Swedish Dairies' Association)
SOU	Statens Offentliga Utredningar (State Committees and Commissions)
SPK	Statens Pris- och Kartellnämnd (National Price and Cartel Office)
	Statens Pris- och Konkurrensverk (National Price and Competition Office)
SS	Sveriges Slakteriförbund (Swedish Meat Marketing Association)
SSR	Sveriges Skogsägareföreningars Riksförbund (Swedish Forest Owners' Association)
Stärkelsen	Sveriges Stärkelseproducenter (Swedish Starch Producers' Association)
SV	Studieförbundet Vuxenskolan (no translation)
TCO	Tjänstemännens Centralorganisation (Central Confederation of Professional Employees)
UK	Lantbrukarnas Ungdomskontakt (LRF Youth Organization)

The Swedish Farmers' Movement and Government Agricultural Policy

Introduction

Research on interest groups has a long tradition within the discipline of political science. The general results of the numerous and varied studies on the subject can be summarized in different ways. One possibility is by type of interest organization studied. Some scholars have dedicated their academic careers to investigations of labor unions, while others have studied employers' associations or farm organizations. Another group has focused its attention on less influential or smaller groups and analyzed their findings in a more abstract theoretical framework, which is not always specific for the study of interest representation or interest organizations.

It is also possible to present the research tradition by summarizing the kinds of interactions that have interested researchers. Almost all studies have dealt, in one way or another, with three different kinds of interactions. One is the relationship between leaders and members of the interest organization. Here the focus is the formulation and representation of the members' interests by the organization, with organizational democracy as a key area of study. The second interaction involves the interest organization with other associations. They may include other interest organizations and political parties. The relationships may be cooperative or antagonistic. Finally, there is the interaction between the interest organization and the state, which may be pluralist, corporatist, or of another type.

Recent research has stressed the importance of studying empirically as well as theoretically how the different relationships affect each other. The insight is the main contribution of the literature on corporatism. The major point is that neither the interest organizations nor its environment (other associations and the state) exist in isolation. It is the research tradition applied in this study.

One of the general subjects currently given much attention in the social science field is the interaction of public and private institutions. To study the issue, scholars have frequently directed their attention to the labor market. Government intervention and the close cooperation between the state and labor market organizations have been the focus. Theories on corporatism are used to analyze the relations between the state and these important, influential organizations and the general political economic development.

The choice of research area is understandable. Yet it is a point of interest that few political scientists have focused attention on a similar set of events and relations. The lack of scholarship in this area has been particularly noticeable in Western Europe, though the situation has changed in recent years. The set of events and relations which have been overlooked are the role of the state in agricultural policy and the role of farmers' organizations as a cooperating and responsible actor in policymaking.

The neglect is puzzling. Many of the defining theoretical aspects of corporatism find their best empirical exemplification in the agricultural sector. Also, active and interventionist state policy has compensated farmers to a higher degree than warranted by their political, economic, and social importance. Chapter 1 offers a general perspective and explanation of this.

The general subject of this book is the representation of the interests of farmers in Sweden. The Federation of Swedish Farmers (*Lantbrukarnas Riksförbund,* LRF) is the center of focus. Its ability to represent its members and safeguard the interests of farmers in the current period of reform in political thinking on agricultural policy is the main subject. The aim is to analyze how this largest and most important interest organization of farmers in Sweden has represented its members in light of the many changes that have affected its strategy, tactics, and position. The constraints on interest articulation from the close ties between the LRF and the state are investigated and analyzed. The advantages and disadvantages of a close relationship with the Center Party for the development of the LRF are discussed.

The book also analyzes the historical situation of the LRF and its forerunners, the National Federation of Agricultural Cooperatives (*Sveriges Lantbruksförbund,* SL) and National Swedish Farmers' Union (*Riksförbundet Landsbygdens Folk,* RLF). An important conclusion of the historical study is that the organizations were greatly affected by state agricultural policy, in particular the policy adopted in the mid-1960s. A primary example is their decision to join together and form the LRF.

As pointed out in Chapter 1 and discussed more fully in Chapters 3 and 6, the LRF is a unique type of farmers' organization as it is a combination of the union and cooperative branches. Also, the farmers' cooperatives are very dominant in the food sector.

A key question that is posed is how well one peak or umbrella farmers' organization can represent the interest of farmers as union members and the cooperative associations as businesses. The issue is a serious one for three reasons. Farmers show a lower solidarity with their cooperatives than with their union. The interests of individual farmers and cooperative businesses do not coincide on all issues. Statistics on membership rates and examples of disagreements are offered in Chapters 3 and 6. Finally, member solidarity is important in times of crisis, the situation facing the LRF today. The attempts by it to promote solidarity and unity among the members in times of change is analyzed in Chapter 6. Its choice of information campaigns for the membership is emphasized, as is the use of metaphors and ideological control by the leadership. Also, the attempts aim to create unity in the heterogeneous corp of members, a crucial issue of the 1980s.

The increased heterogenization of farmers is a consequence of state agricultural policy. Among other things, it has led to the formation of guilds that represent specific interests of groups of farmers within the LRF organization. Another issue focused on in Chapter 6 is the situation facing the cooperative associations, which has changed as a result of development of the Swedish economy and agricultural regulation. Chapter 5 discusses the legal parameters of cooperatives in Sweden. Unlike certain other countries, cooperatives are not exempt from antitrust laws. Some of the basic principles of the Swedish farmers' cooperatives have been found to be in violation of the laws protecting market competition.

A consideration of importance in this regard is the ability of the LRF to govern the cooperative associations. The federal structure of the LRF has generally discouraged more centralized control. Yet it seems that centralization is a key solution to some of the more general problems facing individual cooperatives. The attempts by the LRF to create an understanding and acceptance for centralization is discussed in Chapter 6.

A third focus of the book is the role that peasant romanticism plays as an argument for safeguarding the interests of farmers. It is interesting to note that the farmers' organizations did not initially formulate the interests of their members in romantic terms. Since 1985 the LRF has used peasant romanticism in its public opinion formation campaigns. The role of myths in the interest representation of farmers is discussed in general terms in Chapter 1. Examples from the Swedish case are examined in Chapter 6.

A considerable portion of the book is devoted to an account of the treatment of agriculture and the interests of farmers in Swedish politics in the 1900s. The role of agriculture in Swedish political development is discussed. The aim is to explain the disproportionate influence of farmers in politics in comparison to their strength in the electorate and economy. Here the Swedish situation does not deviate much from the one found in other countries, which can be generally described in the following way.

Unlike the interest organizations of labor, which are or were forced to fight forcefully for the recognition and political legitimation of the interests of workers, farmers' organizations in many countries met with sympathy from politicians and the state. Their interests were seen as unquestionably legitimate and equated with national sovereignty. It is noteworthy that this attitude is changing gradually. It may be the case that the situation is reversing itself, as farmers' organizations must argue now for the preservation of their position while other groups, such as consumers and wage earners, find the state more sympathetic to their cause. An important question addressed in this book is the ability of the LRF to adapt to the waning exceptionalism of agriculture.

Chapters 2 and 4 analyze the changes that agriculture and state agricultural policy have undergone since the early 1900s. The findings may be summarized in seven general points.

First, agriculture has been considered a special policy area. Exceptionalism characterizes this position. It reflects the idea that agriculture as an occupation should be considered as different from other enterprises. Now agriculture is compared increasingly to other policy sectors. Policymakers consider the issues motivating agricultural exceptionalism to be less important today. They include self-sufficiency and war preparedness, two of the main goals of agricultural policy in Sweden.

The policy domain of agriculture has undergone expansion. Chapter 1 presents the idea of policy expansion in general terms. An important example of changing policy concern is the political debate on deregulation, costs of surplus production, consumer demands, and the environment of agriculture.

As stressed in the chapters, other issues have helped agriculture retain its privileged position in the past. Concerns for rural decay have historically been equated with the farm problem. It is pointed out in Chapter 7 that the association of regional policy with agricultural policy may be broken in the near future. The rural interest can no longer be equated with the farm interest. An important question for further analysis is how the LRF will cope with this situation, as it has been able to utilize the attachment of Swedes to their countryside to benefit its members.

The second major point is the creation of a countervailing power in the agricultural subgovernment by the state. In particular, the social democrats were instrumental here. Chapters 2 and 4 discuss how and why the state granted the representatives of the consumers the role of a countervailing power. It signifies that consumers participate in the regulation of agriculture, price-setting negotiations, and policymaking forums. Price, market, and other types of regulation is a general topic of Chapter 5.

The chapters discuss the broadening of the definition of a representative of the consumer. Whereas producers are strictly defined as farmers, the concept of consumers now includes all interest organizations that represent members who buy agricultural commodities: labor unions and a large portion of the food industry. It is noteworthy that the LRF's food processing cooperatives are not defined as consumer representatives, while trade associations for the private companies are included in the definition.

The ability of the LRF to adapt to the changes in the power balance within the subgovernment is addressed in several ways. Examples include its disputes with the state and the consumer representatives and the creation of common or shared interests with the consumers. Another example is the decision by the farmers' cooperatives to focus more on consumer demands and market signals, which reflects the increasing association of market mechanisms with consumerism by political actors, in particular the Social Democratic Party.

The next two points are party politics generally. They are the topics of Chapters 2 and 4 and concern the declining importance of farmers in electoral politics: the decrease in politicians from the countryside in the Riksdag, the impact of urbanization and industrialization, and the reliance of the political parties on the farm vote.

The evolution of the Agrarian Party into the Center Party has had a significant impact on agricultural policymaking. As implied by its name, the Agrarian Party represented the interests of farmers in the parliamentary process. The close association between the party and the farmers' organizations is illuminated in Chapter 3. Examples of how farmers benefited from the party are taken up in Chapter 2.

The party changed its name in the mid-1950s. Though it is still associated with farmers, the Center Party is interested in presenting itself as an environmental party for urbanites. It has evolved from a political channel for the interests of farmers. The discussion on this point summarizes recent research on the attitude of MPs from the party toward farmers. The development of the Center Party has important implications for the farmers' movement. No longer can the LRF rely always on the party to channel its interests into politics.

The position of the social democrats on the agricultural issue has changed dramatically over the years. This is the fourth general point. The interests of consumers are its primary concern now. As mentioned earlier, the interests of consumers now tend to be equated with the competitive market and deregulation. The development is reinforced by its alliance with the two largest peak union organizations for blue- and white-collar employees, the Confederation of Swedish Trade Unions (*Landsorganisationen i Sverige,* LO) and Central Confederation of Professional Employees (*Tjänstemännens Centralorganisation,* TCO). Also, the party does not seem to view the agricultural issue as essential for political cooperation with the Center Party, which signifies a change in attitude from the past.

As a consequence, the LRF cannot rely on the relationship between the Center and Social Democratic Parties for goodwill. Only recently has the LRF chosen to face the situation and decided to confront the problem from a position of independence. An important research question is how the LRF will compensate for the lack of support from the government.

The entrance of other issues into the agricultural policy arena and other organized interests into the subgovernment is the fifth point. The significance of the change for the LRF is that it must compete with other issues and interest organizations for policy attention. The consequence for agricultural policy strictly defined is that the government must consider what financial priority to give the other issues. The study has found that the budget for agricultural policy has tightened, and money is allocated to other issues that are given a higher priority. It has also been rumored in political circles that the ministry of agriculture may undergo amalgamation with other ministries. The question is whether this would be the deathblow to agricultural exceptionalism.

The sixth point concerns agricultural corporatism. It is a subject of all the chapters, but Chapters 5, 6 and 7 deal with it specifically. Both the LRF and its forerunners have participated in the process of agricultural policymaking and implementation. A reciprocal relationship with the state developed for two basic reasons. The state relies on the expertise of the interest organizations for its street-level agricultural bureaucracies. It needs the support of individual farmers, the policy takers, for successful implementation of policy. The relationship is termed the corporatist exchange in Chapter 5.

The organizations assumed responsibility for policy implementation as one task in their representation of the interests of their members. The fusion of interest representation and policy implementation has affected the scope of organizational activities. It has made the interest organizations more flexible to changes and ready to assume a problem solving role that goes beyond the representation of members.

Frequently, the interest organizations have accepted policy solutions which cannot be said to be beneficial to the members or the organization in the short run. An important prerequisite for the corporatist exchange for the LRF is its ability to convince its own members that the choices are the right ones. Chapters 6 and 7 discuss how the LRF performs the task. The role of leadership and incentives for membership in this process are main topics here.

A main conclusion is that effective and democratic organizational leadership is particularly important in times of crisis. The Swedish style of leadership is discussed, a central component of which is flexibility in decisionmaking situations and awareness of the urgency of the agenda problems. It means that leaders refrain from formulating adamant demands and willingly discuss policy options. At times it may seem that the leadership has not taken a clear position on issues of crucial importance to the membership. Yet the leadership of any association is the key actor in reaching a compromise between the demands of the membership and the political system.

The final general conclusion of the study involves the attempts by the social democratic government to end agricultural exceptionalism through redefinition of the policy problems and reformulation of the ground rules for policymaking. The new problems are deregulation and the environment. The new rules involve the representation of interest and setting for policymaking.

It is intriguing that international opinion is allowed to play a role in the agricultural policymaking of Sweden. The efforts of the General Agreement on Tariffs and Trade (GATT) and European Free Trade Association (EFTA) to change the direction of agricultural policy in western industrialized countries and the conclusions reached by the Organization for Economic Cooperation and Development (OECD) are given top priority in the current Swedish debate. The opinions of the international organizations are an interest that is represented in policymaking, as their results form a basis for the work group that is formulating the reform of Swedish agricultural policy.

The decision by the government to commission the ministry of agriculture to appoint a ministerial work group to study agricultural and food policy that is composed of Members of Parliament (MPs) from the different political parties is noteworthy. The MPs are members of the Standing Committee on Agriculture in the Riksdag. Chapter 7 discusses the status of the work group and suggests that the government has chosen this method for interaction in order to create a more favorable setting for the demise of agricultural exceptionalism.

The research project was generously funded by the Swedish Council for the Planning and Coordination of Research (*Forskningsrådsnämnden*). Olof Ruin encouraged me to undertake the investigation. The LRF was very helpful and open in supplying information. Two LRF officials, Lars Idermark and Mats Denninger, deserve special mention. Barry Holmström read the manuscript carefully and offered valuable comments. The Swedish Council for Research in the Humanities and Social Sciences provided funding for the camera-ready production of the manuscript.

1

Every Nation's Heartland

FARMERS AND POLITICS

Politics and weather are two concerns that constantly weigh heavy on the minds of farmers. Over the years, farmers have tried to cope with climatic problems by adapting to them. When technology has not offered a viable alternative, mother nature has reigned over the crops. Possibly as a result of the unruliness of nature and definitely due to the essential importance of food supply for national well-being, politics has come to play a primary role in agriculture. This is even the case in countries where state intervention usually is at a minimum.

It is not surprising then that farmers attempt to influence politics. Whereas farmers have learned to accept weather for what it is, they have not acquiesced to politics. As a group, farmers have been more successful in achieving their goals than other groups of comparative size. Comprising less than 5 percent of the population in most of the advanced industrialized nations, farmers continue to be a powerful political voice. But their influence is diminishing.

Several reasons can be given for the political success of farmers, but three are basic. First, successful agriculture was traditionally an important source of national wealth. Agricultural products could be traded for goods that could not be produced domestically. Even individual wealth was and continues to be measured by food habits.

Sufficient food supplies were also necessary for industrialization and urbanization, with both occurring as it became feasible for fewer people to produce more food. At present, a prosperous agricultural sector continues to be regarded as a national necessity, and it is viewed with a sense of national pride. The primacy of agriculture for national livelihood has time and again been underscored.

A second reason for the political influence of farmers is the rural myth, peasant romanticism, and the agrarian tradition. The terms refer to the idealization of farmers and farming as a way of life. The idealization has occurred in different countries and historical epochs. Though formulated in many ways, its main message is reverence of agriculture. One important aspect is the connection between farming and nationalism. In this view, farmers are considered as more patriotic than other groups in society. Agrarian patriotism has been explained in terms of the more homogeneous nature of the rural population, its lack of contact with foreign countries and peoples, and the limited geographic and social mobility of farmers as a group. Xenophobia, racism, and fascism have been associated with agrarian patriotism.

Peasant romantics place farmers above other social groups. Rural society is seen as superior in national spirit on moral and material grounds. It is said to preserve national character. The family farm plays an important role here and is exalted for its decisive contribution. Families working together on farms symbolize rural values such as attachment to the land, respect for the stability arising from property ownership, good family life, high valuation of work, rigid morality, individualism, and economic self-sufficiency.

Also, the rural myth furthers the view that life in the countryside is superior to life in the cities. Rural pride and an explicit association of ruralism with nationalism can be found in most industrial countries. Two common expressions in Swedish history are farmers as the "backbone of the country" (*landets ryggrad*) and "trunk of the societal tree" (*stamman i samhällsträdet*) (Torstendal 1969, 27). They contrasted sharply with urbanism, which was depicted as superfluous and leading to national and individual demise. An American farmers' organization proclaimed rural life as producing the best citizens of democracy. The disappearance of the family farm is said to "be felt in a loss of American stamina, tone, and nerve" (Crampton 1965, 27). The metaphor is ruralism as the *body* of politics.

Farmers joining together into associations is the third essential reason for their influence in politics. Academic disciplines as different as political science and economics agree on this conclusion. Both disciplines view political influence as the ability of farmers to succeed in generally getting what they want out of politics, that is, favorable legislation and subsidies. Political scientists have been particularly profusive in investigating the political power of organized farmers, the main thrust being the special political privileges accruing agriculture.

A number of theses have been raised as explanations for this "exceptionalism" (Grant 1987, 157). Unfortunately few authors deal directly or systematically with the impact of the rural myth on modern politics. Instead, pressure group politics has been the subject given most attention. How farmers become and stay organized and how organized farmers influence public policy are the two related questions highlighted in political science research.

When compared to other special interest/government relations, the one between farmers and the state in western democracies shows a number of revealing similarities. Strong farmers' organizations are usually present. Yet the pattern of relations differs, as does the number of organizations involved in it. Three essential factors characterize the political contact pattern: whether a particular political party identifies its cause with the special interest of farmers, whether lobbying is a primary means of communication with and channel of influence, and whether farmer organizations are confronted with rival interest organizations when agricultural policy is made.

In contrast to all western countries, farmers in the Nordic countries have been able to rely on durable and electorally significant agrarian parties to represent their interests. Research has shown that agrarian parties developed in countries where: a large part of the rural population was family-based, self-sufficient and non-feudal; cities and industrial centers were weak when suffrage became more general; a major cultural cleavage involved conflicts between the cities and countryside; and the Catholic church was not important (Lipset and Rokkan 1977). The agricultural interest has generally been accommodated by the non-socialist parties in the majority of western European countries (Urwin 1980). In the United States, the Democratic and Republican Parties both represent the agricultural interest. What makes interest representation in America particularly different is that the parties do not support the same agricultural commodities.

The Danish, Norwegian, and Swedish parties played decisive historic roles. This occurred in the 1930s when they agreed to support social democratic minority governments. The general trade-off in all three countries involved the social democrats going against their position on free trade and accepting agricultural protectionism and subsidies along with other benefits to farmers. In return, the agrarian parties promised to give their parliamentary support for social democratic proposals on unemployment relief and crisis policy. The agreement has been claimed as a unique "historical compromise" since it provided a stable basis for government and the development of the Scandinavian welfare state (Esping-Andersen 1985; Korpi 1978).

Though historical compromises of such grandeur are limited to Scandinavia, politicians have taken the opportunity to use agricultural policy to promote welfare aims. The food stamp program in the United States was, for example, logrolled through the agricultural committees of Congress (Hansen 1987).

Lobbying is one of the characteristic ways by which interest organizations exercise political influence in the United States. Yet as a means for goal attainment, it is inferior to an intimate interest organization/political party relationship and, as discussed below, sponsored pluralism and corporatism. Lobbying implies that the organized interests are not part of the government process and that their influence in politics may be rather sporadic.

Generally lobbying means that interest organizations try to exercise influence by contacting government officials directly in order to inform them about their views, testifying at committee hearings when a bill is before the legislature as well as using the civil court system for goal attainment. Though the farm lobby is strong, the exceptional political influence organized farmers have been able to exercise over agricultural policy comes from another source.

Regardless of nation, the political influence of farmers comes in large degree from close relations between farmers' organizations and policymaking branches of government. Contacts with ministries of agriculture and legislative committees are the most important. This close cooperation has been termed a "partnership between the state and the farmer" and is characterized by reciprocal cooperation.

Two general theoretical approaches, sponsored pluralism and corporatism, are useful in analyzing the partnership. Both approaches have created a great deal of academic controversy over how they should be defined and what actually distinguishes them from each other (*see* Young 1987 for a summary). They are similar "in that in both policy networks associations are incorporated into the processes of policy formulation and implementation and may be delegated certain state powers" (Atkinson and Coleman 1985, 25).

For the purposes of this book, the following distinctions are made. *Sponsored pluralism* refers to a situation where government policymaking does not incorporate two countervailing interest organizations. For agriculture, this means that interest organizations representing consumers do not confront the farmers' organizations in their negotiations with the state. The distinguishing characteristic of *corporatism* is the incorporation of antagonistic interests that are given legitimacy by the state to compete in a policymaking setting (Cox 1988; von Beyme 1983).

The number of interest organizations involved in the making of agricultural policy is an important point. Later in this chapter, changes in agricultural policy that have occurred in different countries and over the past decades are discussed. One interesting development that helps explain the changes is the introduction of other interest organizations or interests or both in the policymaking process.

Certain well-known concepts in political science are used to describe agricultural policy. One is the *iron triangle.* Here politics is studied through analysis of policy areas or sectors rather than simply political institutions. A policy sector involves people who are interested in a particular public policy issue and who come from legislative committees or subcommittees, administrative agencies within a ministry, and interest organizations.

The iron triangle metaphor is a good one. It conjures up an image of impenetrability in that politicians, administrators, and leaders of interest organizations outside the system find it difficult to participate in policymaking. Political actors inside the policy triangle experience a sense of security since job changes can occur from within and do not break up the close working relationships. Studies on iron triangles have shown that they work well for those involved. Participants share a common view of policy and try to reach a consensus that is mutually beneficial for all concerned. Yet outsiders wishing to influence policy may find that their viewpoints are considered uninformed and are ignored.

The concept of an iron triangle has shed much light on agricultural policymaking. Even when not used directly, it offers an interesting perspective for summarizing research results. For example, a number of studies have shown how the iron triangle of agriculture has set the agenda for public debates on agricultural policy. One West German investigation reports that organized agriculture was able to remove the issue of high farm prices from the political agenda principally by channelling it into a nondecisionmaking process (Andrlik 1981). A comparative study of agricultural policies and politics in Britain and the United States discusses the role of American farmers' organizations and concludes that they have "a golden opportunity to set the terms of debate on other people's proposals" (Wilson 1977, 99). Policymaking in an iron triangle is one characteristic of sponsored pluralism.

Other researchers use the iron triangle concept to analyze change in agricultural policymaking. They stress that the iron triangle of agriculture has corroded. The American farm bloc began to break up when a proliferation of the farmers' interests took place. In some countries, other interests successfully demanded a say in agricultural policy. The role played by labor market organizations in Sweden changed the emphasis of Swedish agricultural policy. In a similar way, urbanization affected American policymaking. The interests of consumers became an integral part of agricultural policymaking in both these countries. Though in other countries, for example Great Britain, consumers are still excluded (Smith 1988; Yale Law Journal 1962).

The term *issue network* has been introduced to analyze policymaking. An issue network is a subsystem which is more open and accessible to a broader range of political actors (Heclo 1979). An issue network is corporatist if it involves the incorporation of interest organizations in policymaking that are countervailing powers and granted legitimacy by the state. All issue networks are not characteristic of corporatism.

ORGANIZED AGRICULTURE

Over the span of one hundred years, farmers have created associations to further their interests. Earlier associations differ from current interest organizations in significant ways. Associations formed at early dates and in a number of countries to help in the transformation of the system of farm production.

The modern Danish agricultural interest organizations were, for example, built upon older associations of this type (Laux 1968). The oldest Swedish farmers' organization dates from the late 1700s and dealt mainly with agricultural information. In a similar way, it aided Swedish farmers and continues to play an active role in distributing information on Swedish agricultural policy. Farmers' cooperatives existed also in the 1800s, as did political parties catering to the interests of farmers.

The economic problems that farmers faced in the first decades of the 1900s gave rise to another kind of association, the modern interest organization. Modern farmers' organizations exist in all western industrialized democracies and were created to promote the economic interests of farmers. Two types of organizations are present in most countries. The first type, *farmers' unions,* seek to strengthen the bargaining power of farmers in much the same way as labor unions promoted the interest of wage earners. *Farmers' cooperatives* aim at providing a stronger and more stable position for individual farmers as businessmen.

National farmers' unions are found in a number of countries. Yet there is a clear contrast between countries where one union dominates and those where more than one union competes for membership and political influence. One union domination is present in the United Kingdom, Sweden, and West Germany while different unions compete in France as well as the United States and two in Denmark and Norway. The Danish union representing medium-sized farmers was created in 1893 and the union of small farmers thirteen years later. The pattern in Norway is similar: the union for larger Norwegian farmers was established in 1896, 17 years before the union for small farmers. The dominant farmers' union in Sweden was formed in 1929 (Ball and Millard 1986; Steen 1981, 1988; Micheletti 1987a; Buksti 1974).

National associations for commodity cooperatives are more unusual. Some tasks performed by cooperatives are included in the activities of the farmers' unions. Farmers' or producer cooperatives which are not quasi-governmental in character are uncommon in the United Kingdom (Grant 1985). One single organization of agricultural cooperatives exists in Japan. The organization differs from the American ones in the massive size of its membership, personnel strength, magnitude of business operations, and range of functions. One author describes it by referring to the organization's ubiquitous presence in the countryside (George 1981).

Unlike the situation in the United States, national commodity cooperatives as well as a united nationwide federation for the different national cooperatives exist in Scandinavia. The national federation for Swedish farmers' cooperatives was established in 1917, its Norwegian and Danish counterparts in 1945 and 1917 respectively.

The Scandinavian countries stand out in terms of organization of farmers' interests as well. Not only were there special agrarian political parties representing the interest of agriculture, as discussed earlier. It can be maintained that the union and cooperative organizations of farmers are generally stronger than in most other countries. The reasons include the smaller number of organizations, the low level of competition among them, their relatively high organizational densities (i.e., higher proportion of potential members as members), their nationwide basis, and their large relative size in comparison with farmers' organizations in other countries (for comparative statistics on organizational density see Ball and Millard 1986, 135ff.).

The Federation of Swedish Farmers (*Lantbrukarnas Riksförbund,* LRF) should be singled out in the general Scandinavian pattern of organizational prominence. The LRF is internationally unique as it represents farmers in their capacity as both businessmen and union members. (One exception is a small organization in Belgium which represents the union and business interests of farmers in much the same way as the LRF. It organizes a much smaller proportion of farmers.) The LRF was created in 1971 when the cooperative branch and the union branch of the farmers' movement decided formally to join forces and establish a united farmers' movement. Its members farm over 75 percent of all arable land, and the LRF organizes about 80 percent of all farmers with holding of 20—50 hectares. (A hectare is about 2.5 acres.)

INTEREST REPRESENTATION BY FARMERS' ORGANIZATIONS

A review of the literature on farmers' unions shows that the unions tend to have three basic goals which they attempt to attain in four ways. The first goal is *equality of agriculture* in relation to blue-collar workers. As discussed in the section on agricultural policy, it is most frequently formulated in terms of a yearly income equal to that of industrial workers. In order to attain the goal, farmers' unions negotiate with cooperatives, other wholesalers, and the government over prices. This is the *collective bargaining task* of farmers' unions. Collective bargaining results have been found to be insufficient in attaining the goal of equality. Therefore farmers' unions press for legislation that supports farm communities, and they attempt to gain public support for their cause. This is the *political task* of farmers' unions.

Another general goal is *maintenance of the family farm*. In significant ways, the goal contradicts the first one, as income equality is more a question of economies of scale. The basic interest represented by the second goal is preservation of traditional farm life, the rural population, and countryside. Farmers' unions reason that they can exercise political influence more successfully when the size of their memberships are larger and when politicians and the general public consider rural values as an important national asset. This is the *mythicizing or symbolic task* of farmers' unions.

The third main goal is *provision of protection*. Farmers' unions provide a number of services to their members. Insurance is the most common and possibly most important service, but others as education, agricultural counseling, loans, and discounts on purchases are offered also. It involves the unions in the *task of selective incentive provision*.

Several tactics are used by farmers' unions to attain their goals. Common ones include campaigns aiming at informing the public on the situation of farmers, lobbying and participation in the policymaking process. More belligerent tactics—strikes, holding and dumping actions, demonstrations, and violence—are more frequent in times of desperation. When the economic situation of farmers is bad and when farmers believe that government is not responding to their problems, militant tactics tend to be used more often. Political culture as well as union militancy play an important role in the choice of tactics. France, the United States, and the Scandinavian countries will serve as contrasting examples.

French farmers used militant tactics in the 1980s, thus ending a period of calm since the early 1960s. The change is explained by the deterioration of the economic situation of the French farmer, the impact of the European

Community's Common Agricultural Policy on French agriculture, and the recent "divorce" between the French government and the most powerful agricultural organization in France (*Féderation Nationale des Syndicats d'Exploitants Agricoles*, FNSEA).

Analysis of the French situation shows that the final explanation is the most important. It signaled that the newly elected socialist government wanted to stop giving the FNSEA priority over other farmers' organizations in agricultural policymaking. It was the FNSEA that excited French farmers to drive their tractors through Paris, insult and throw eggs at the minister of agriculture, and engage generally in contentious activities. Dumping actions and violence are two other tactics that were employed (Keeler 1987).

A somewhat different set of events which culminated in much the same way occurred in the United States. Lobbying tactics were developed when organized agriculture began to sense that government did not attend to the needs of farmers as it had in the past, due to the break up of the bipartisan farm bloc. The farm lobby was initially very successful in furthering the interests of agriculture, but it lost influence over agricultural policymaking as society changed. Conflict among farmers' organizations also played an important role.

By the late 1970s, the situation had deteriorated to the point that U.S. farmers, following the tactics used by other disgruntled groups in society, took to the streets. An estimated 100,000 tractors participated in "tractorcades" throughout the country. One slogan adapted the anti-Vietnam War cheer "hell no, we won't go" to suit the protests of farmers: it became "hell no, we won't grow." Delivery strikes occurred but were less successful. Individual farmers belonging to a more protest-oriented union even lobbied Congress in the first three months of 1978, but their efforts did not have the desired impact (Fite 1981). Today there are over one thousand lobby groups serving the interests of agriculture in the United States.

Neither is conflict foreign to the Scandinavian countries. "Milk as a weapon" is the title of a study of the Norwegian farmers' movement in the mid-1940s. The aim of the milk strike was to secure income equality (Roald 1977). Supported by the local branch of the farmers' union, farmers in the Swedish province of Dalarna refused to deliver milk in 1942; they also wanted to improve their economic situation (Thullberg 1988). Danish farmers have reacted in the same way. Backed by their union in the early 1960s, they instituted a production stop, an action which they hoped would force the government to acquiesce to their demands (Laux 1968).

Strikes or holding actions have historically been moderately successful in Scandinavia. It can be questioned whether more militant action may not become common in the future. Will farmers in other countries face a similar situation as their American or French brethren? The answer is affirmative for Sweden. Demonstrations, dumping actions, embargoes, and tractorcades have occurred recently. Swedish farmers are becoming more militant.

Another way of assessing the influence of farmers is the breadth of interests represented by the organizations. Here the subject is the legitimacy of the farmers' organization in the eyes of its members, other interest organizations, the public, politicians, and government officials. The issue is discussed with reference to the Swedish case.

The LRF is the most encompassing of all farmers' organizations. Not only does it include the interests of farmers as union members and businessmen. Its organizational density is so high that the other minor farmers' unions are not considered to have any real legitimacy. This differs from the experience of most other western countries.

Organizational strength of this kind implies generally that the LRF must aggregate the different demands placed by its heterogeneous membership internally before it pressures government for support and policy changes. Whether the process of internal aggregation satisfies the membership's demands and whether it contributes positively to the agricultural policymaking process are two important issues focused on in this book (Olson 1982; Micheletti 1987b). The ability of competing farmers' unions to unite on issues and seek the support of other interest organizations is an attribute for the exercise of influence. The situation in the United States is illuminating.

It is convincing to analyze the political plight of U.S. farmers by focusing on the lines of conflict both within the agricultural sector and among the farmers' unions. Inter-crop conflicts escalated as agriculture underwent change. A proliferation of interests which farmers' unions must represent occurred, and it proved difficult for the organizations to find a common point of departure for united action (Wilson 1977; Yale Law Journal 1962). As the different unions represent only a fraction of the heterogeneous farming population, they often concentrate on articulating a narrow interest that does not take other farming interests into account. The same pattern is found among the national commodity associations in the United States, which can be characterized as lobbying and trade organizations.

Conflict becomes inevitable. Attempts to formulate a common farmers' interest in 1957 failed, for example. Interorganizational disagreement frustrated politicians who represented rural areas: Farmers' unions could not agree on the role of government in the farm problem. Whether farmers should ally with business or labor was another area of disagreement. The application of the goal income equality to all farm commodities or only a few was a third. The unions continue to disagree over the use of tactics.

Some argue for more militancy, while others emphasize the primacy of lobbying and public campaigns.

Many of the disparities are explained by political sympathy. Another explanatory factor is the divergent structure of U.S. farms. Certain U.S. farmers' unions have identified historically with socialist parties. Others support either the right- or left-wing of the Democratic Party while the largest union identifies with the Republican Party (Fite 1981; Lipset 1971; Browne 1982). The differences led to a weakening of the political strength of farmers, a sensitive development as consumers were gaining in influence.

Political sentiment plays a role in the Norwegian farmers' movement, with the majority of larger farmers sympathizing with the Center Party, while smaller farmers who supplement their incomes with part-time work in blue-collar employment overwhelmingly support the social democrats. The two groups are also organized in different farmers' unions and tend to disagree on certain issues of Norwegian agricultural policy (Sundsbo 1971; Gåsdal 1985). A similar development is present in Denmark.

Exactly how important the organizational differences are for the general influence of the farmers' movement is an interesting question that is not answered easily. It seems reasonable to assume that a conflicting pattern of relations among farmers' organizations will be damaging in times of agricultural crisis. The present debate in western democracies on the reform of agriculture may prove to be a good test of this assumption.

Another way for farmers' unions to exercise influence is by using the rural myth to rally support among the members, general public, and politicians. When compared to farmers' organizations in other countries, it is quite surprising that the U.S. organizations have not been able to use the myth to convince their members to agree on a common farmers' interest. Other organizations, such as the LRF and rural associations in France, have used the dichotomy of city versus countryside to create a sense of solidarity among the members in order to hold their memberships together (Berger 1969, 1972; Peterson 1985).

Evocation of the rural myth in contacts with government has proven to be successful in rallying political and public support for the farm problem in several countries. It has been a good tactic for obtaining external support. Focusing attention on the problems experienced by farm families has instilled politicians and the public with a sense of guilt and perpetuated an obsolete conception of farm life. Whether farmers' organizations will be able to gain support by mythicizing about their situation in the future is question that deserves further study.

GOALS OF AGRICULTURAL POLICY

Problems in the agricultural sector were given political priority in many western countries in the 1930s. Politicians realized that if the problems were not solved they would disrupt politics and the economy. It is noteworthy that the western nations formulated policies which sought to solve these problems in similar ways. The measures that were proposed over 50 years ago form the basis of current agricultural policy. This section outlines the agricultural policy which has developed and evaluates how well it has served farmers and the nation as a whole.

Modern agricultural policy includes one or more of the following four major objectives. What is especially interesting about the objectives is that they highlight how agricultural policy has come to solve problems which are really not farm problems per se. The first problem involves national defense. Probably one of the main concern of politicians of the past was self-sufficiency in the production of food, a problem related directly to the vulnerability of nations in times of war. Due to its crucial role here, farming has been equated with the national interest. A second basic goal attempts to solve the problem of economic dependence on other nations, an objective also related to national autonomy, that is, reduction of balance-of-payments difficulties.

The final two objectives are of more recent origin and reflect the welfare policy concerns of industrialized nations. One is the economic situation of the farm population. In many countries, agricultural policy in the 1930s became a social policy for farmers and an employment and a regional policy for rural communities. The other aims to improve the yearly income of farmers through various kinds of renumeration, subsidies to stabilize the prices of agricultural products, and measures to expand the employment opportunities in the rural areas.

As society changed, politicians and government officials became concerned about the effects of agricultural policy on consumers. A fourth goal, the availability of food at reasonable prices, was formulated (Johnson 1973; OECD 1987, 1988b). A final general objective is evolving now in post-industrial countries. It reflects concern for the detrimental effects of agricultural chemicals on the environment and food quality.

Students of agricultural policy stress that every industrial democracy protects and subsidizes its farmers. Agricultural policy is generally considered to be client politics, as the interests of agriculture hold a prominant position in policymaking. National differences in general political ideology and culture have not affected the formulation of agricultural policy significantly. Even countries which tend to keep government at a minimum and to follow free trade policy protect their domestic agricultural sector.

The *exceptionalism* with which government treats agriculture is evident in various ways. Unlike other branches of industry, shipping or steel for example, a vast network in the public administration and government exists to deal specificially with the problems of agriculture. Ministries of agriculture are common in most countries as are standing legislative committees, which formulate the specific proposals, and government agencies, which administer the policy goals and measures.

Subsidiary goals are often formulated to complement the basic ones. In turn, their implementation leads to the creation of other administrative units. For instance, it is common that countries decide to grant special loans to farmers, which bring about the establishment of public credit institutions. In many countries, the legislative branch of government has also decided that farmers should be provided with free advisory or extension services and that agricultural research should be supported publicly. National and regional boards and agricultural universities and colleges were created as a result. Few if any other policy areas are distinguished by their own public university system. In some nations, the agricultural public administrative network is one of the largest, if not the largest, policy sector in terms of size and scope of activity.

It is also possible to generalize about the implications of agricultural policy. A number of the goals have generally been attained. Most western nations have achieved self-sufficiency and do not need to import basic food commodities, thus eliminating problems of agricultural trade deficit. Most farmers are in better economic shape now. The consumer goal of agricultural policy is more difficult to evaluate, but food supply has stabilized over the years.

UNINTENDED CONSEQUENCES OF POLICYMAKING

Western agricultural policy is characterized by a significant number of unintended consequences. Policy measures devised to ensure self-sufficiency and reasonable incomes for farm families are creating sizeable commodity surpluses. Agricultural subsidies have generally tended to favor larger farms over the smaller family farms once politically targeted as in need of help. The costs incurred from perpetuating the system of agricultural support are frequently paid by higher food prices and taxes.

Many experts, including international organizations as the OECD and GATT, claim that world agriculture is in disarray since agricultural policy distorts prices and trade, and imposes large costs upon consumers and taxpayers. It also encourages an uneconomic expansion of farm output in the industrial countries while developing countries continue to struggle with the problem of self-sufficiency (GATT 1988; OECD 1988b; Johnson 1973; Gardner 1983). The costs of public support to agriculture are high indeed. By the late 1950s, agricultural commodity programs were the third largest expense after defense and debt service in the United States (Gardner 1983; Hansen 1987, 11). Table 1.1 illustrates the general costs in a number of OECD countries. On an average, Sweden ranks second highest on public support to agriculture.

Table 1.1
Public Economic Support of Agriculture in Countries Belonging to the OECD, as Percentage of the Price Farmers Receive for Commodities, Average for 1984—86

						Sweden		
Commodity	Australia	USA	EC	Japan	New Zealand	1984	1985	1986
Wheat	13	44	36	98	13	22	21	47
Feeding Grain	8	30	26	98	9	23	17	40
Beef, Veal	8	9	53	55	11	52	54	53
Milk	52	66	56	82	14	60	72	87
Pork	4	6	6	40	13	31	29	43
Poultry	4	10	27	16	21	70	68	69
Sugar	21	76	75	72	*	56	68	67
Meat	15	26	41	46	23	45	52	59
Vegetables	13	31	38	86	11	30	28	39
All commodities	15	28	40	69	23	39	41	55

Source: Adapted from Ds 1988: 54, 32.
Public economic support is defined as all categories of economic support given to agriculture that is financed through the national budget or by consumer food prices.
* Not available.

Many writers attribute a number of the problems facing agriculture to the practice of *administered protectionism*. The term refers to two important aspects of agricultural policy: how it is implemented and where it is decided. Western industrialized nations protect domestic agriculture by using distinctive policy instruments. Border protectionism regulates agricultural trade and is directed at both prices and quantity. Examples include: import levies and export restrictions which aim at stabilizing domestic prices, tariffs that raise the price of imported goods, export subsidies which reimburse farmers for exporting goods at prices lower than domestic ones, import quotas, and health standards.

Also, western nations intervene in the domestic markets for agricultural goods by regulating the market. Measures include production quotas which restrict the amount of a commodity that a farmer is allowed to grow or sell at a favorable rate, input or acreage controls, intervention prices (i.e., the prices that the government is willing to pay for agricultural products), and consumer subsidies (e.g., the U.S. food stamp program and free school milk programs) which offer compensation for food prices (Winters 1986).

The administered protectionism of agriculture is implemented by controlling the market for agricultural goods. But the control is not efficient. Scholars agree that protectionism leads to higher prices, large farm bureaucracies, and market disturbances. Investigations on why protectionism prevails inevitably focus on agricultural policymaking. Emphasis is put on the strength of farm lobbies, the goals of agricultural policy, and the inertia that characterizes the reaction of public administrative units to change.

Here attention is focused on the low profile of agricultural policy. A distinction is made between *low* and *high policy tracks*. Agricultural policy has traditionally been on a low policy track. When compared to high tracks, policymaking in a low one is technocratic: it minimizes the political but not the economic costs of making decisions, it provides a way of avoiding conflict among opposing groups of interests, it favors the interests that are effectively organized and represented in the policymaking setting, it disregards the national interest in policy deliberation by concentrating on parochial ones, and most important, it rests upon public unawareness or political insignificance of the policymaking (Finger, Hall, and Nelson 1982).

Participants in low policy tracks tend to be isolated from outside political influence and conflict. It is probable in certain polities that the participating politicians and public administrators become dependent on the interest organizations and their leaders for their help in administering policy programs. As a consequence, the government loses its autonomy in agricultural policymaking (Lehmbruch 1987). Policymaking inside iron triangles usually involves low track decisions.

Which interests this kind of decisionmaking benefits is a question that deserves close study. Two scholars of the English case offer a definite answer. They state that the close partnership between the farmers' organizations and the state disfavors the interest of farmers since the organizational leaders become prisoners of the state. They claim that the partnership has induced leaders to concentrate entirely on a governmental solution to the problem of farmers (Self and Storing 1962, 122).

Not only does corporatism and sponsored pluralism generally tend to endanger public policy by limiting the range of plausible options (Wilson

1977, 166). The bargaining it involves has been shown to restrict how interest organizations choose or are allowed to react to the demands of their members. The general pattern is that organizational leaders seek to be granted an important and permanent role in policymaking and seek to pacify their members, whose demands might in fact require more radical action.

A study of the British National Farmers' Union (NFU) found that the government at times had made a decision on a particular point of agricultural policy before consulting the NFU. The NFU was placed in a difficult situation. The extreme positions are clear: either it denounces the decision or it tries to convince its membership that the decision is fair (Wilson 1977).

WANING POWER OF ORGANIZED AGRICULTURE

Urbanization and industrialization are the two important changes of the 1900s that have greatly affected agriculture. The developments altered the demography of western democratic nations. Large numbers of people moved from the countryside into the cities in order to seek work in industry. In turn, the rural areas were depopulated and farming became an occupation of a minority that still decreases in number.

The general effect is that other groups use pressure politics to argue for an agricultural policy that takes the needs of other groups in society into consideration. The evolution from an agricultural iron triangle to an issue network in the United States meant that "(t)he nation's agricultural policy arena, long the exclusive domain of farm interests, had, by the late 1970s, expanded to include consumer groups, labor unions, and the 'hunger lobby'" (Cigler and Hansen 1983, 86). In countries where agricultural policy was decided in a different type of setting, newer social groups pressured government to consider their demands.

The general structure of political bodies was affected by urbanization and industrialization as well. Political parties catering to urban and industrial needs were born. Other parties either lost some of their strength due to changes, evolved to include new groups of voters, or simply died out. Legislative reapportionment took place, and electoral districts were redrawn with a major consequence being an increase in the number of representatives from urban areas and a decrease in those from the countryside. The change had a great impact on the representation of the interest of agriculture in the U.S. Congress and the strength of certain political parties in other political systems as well.

Other changes affected the political position of agriculture. Reforms created new legislative settings, for example when the system of estates was abolished and when unicameralism replaced bicameralism. The reforms tended to limit the political power of the rural population. Also, amalgamation of small municipalities into rationalized larger government units affected the political clout of farmers. One scholar of U.S. agricultural policy summarizes the effect of the changes in the United States with the phrase an "ever-decreasing grandstand" for farmers, that is, a decline in their historical and seemingly inherent base of power in politics (Hansen 1987). A similar assessment of the current situation of farmers in other countries can be made.

When seen together, the changes in attitude and developments in society signal that agriculture is losing its status of exceptionalism. Government financing of agriculture is a debated topic in the countries surveyed briefly in this chapter and inter- or super-national forums as the European Community (EC), Organization for Economic Cooperation and Development (OECD), and the negotiations on the General Agreement on Tariffs and Trade (GATT). A reevaluation of the role that farmers play in national defense and the importance of a viable rural community are questions that many nations are dealing with now.

One of the most important reasons for the reevaluation is the immense quantities of surplus products and the expense that they incur for national budgets, taxpayers, farmers, and consumers. Surplus production cannot, as in the past, be sold on the world market for profitable or satisfactory prices. Neither can industrialized countries consume more food. Economic decline in many third world countries has meant that they cannot even afford to buy more food at lower international price. Yet farmers are encouraged by the complicated and contradictory system of financial support to produce as much as possible. The general direction of agricultural policy and the technical advances in agriculture have encouraged this development. Today experts generally agree that this is unwise and argue for a reform of agricultural policy.

A question of interest for this book is the role that the farmers' organizations will play in the reevaluation and change of agricultural policy. It is evident that the economic problems facing individual farmers and the public criticism of the high costs of agricultural policy and food put pressure on farmers' organizations and their leaders.

The membership may demand that the leaders do not compromise in their negotiations with government bodies over prices, subsidies, and changes in agricultural policy. They may want their leaders to represent their interests in a more adamant way. The government, concerned interest organizations, and the public may demand that the farmers' organizations accept policy changes. It implies that they want the organizations to be able to adapt to change, convince their memberships that reform is inevitable, and accept that it is wiser for them to be open-minded about a reform of agricultural policy.

How well farmers' organizations will be able to satisfy the demands of their members and of the political context depends on a number of factors. One is the role of government in the agricultural sector in the future. The interesting question is how the government views its policymaking role. Since agricultural policy encouraged the present situation, it is important to analyze whether government spokespeople consider themselves responsible for helping to reform policy in such a way that does not lead to great political and economic harm to individual farmers as well as their organizations. Will governments assume responsibility for their past actions? An issue of importance here is how responsive government is to policy demands that are voiced by countervailing powers in the agricultural sector.

Another factor is the proclivity of the farmers' organizations to change. Can farmers' organizations reformulate the interests they represent in order to attract new groups of members while still supporting the group of farmers who form the core of the membership? Is it possible for them to expand the sphere of their interest representation to include related interests? Examples of related issues and interests that may possibly be included are the countryside, consumers, quality of food, aid to countries that are not self-sufficient in food production, the environment, the poor, the health issue, and the conservation movement. Expansion of the sphere of interest representation through the addition of members with new interests will give organizations greater flexibility in formulating policy proposals. It occurs since internal interest aggregation of such diversified interests must take place before the leaders express their proposals in an external setting.

Solidarity among different groups of members and between the members and the leadership is a third important factor for the potential political influence of the organizations. Change in agricultural policy requires the responsiveness of organizational leaders and members. When politicians and government officials believe that the viewpoints of the organizations are supported by the membership, they will give them greater consideration. Otherwise the reform policy may fail, as members refuse to obey the new regulations.

Yet if it is believed that the leaders do not have the support of the membership, their opposition to new proposals may be viewed as inconsequential, as the organizations are not viewed as credible representatives of the interests of farmers. The government may decide to proceed with policy reform. It is important that support for the organizational leaders is manifested by the members.

Also, solidarity within the organization is necessary if the leaders are to succeed in articulating an organizational proposal supported by the entire membership. When different groups of members cannot agree on policy, it will prove difficult for the leadership to present a concrete proposal and argue convincingly for its acceptance. Here solidarity should be taken to mean either cooperation among the different member groups so that a compromise proposal can be articulated or transformation of their demands into a new goal for the organization (Micheletti 1987b).

The importance of member solidarity and support for the leadership should not be underrated in times of change. A U.S. study shows that members of Congress questioned the legitimacy of one farmers' organization because they did not believe the leadership had the support of its members. Rather it was assessed that member solidarity in the organization was based on the provision of selective incentives (Cigler and Hansen 1983).

The character of interest organizations and social movements that are not farmers' organizations but that have demanded a say in agricultural policymaking is a final type of pressure put on farmers' organizations. The main issue is whether their presence has changed the role and influence of farmers' organizations.

In the first place, it is necessary to study whether the other organizations are a *countervailing power* in agricultural policymaking. The concept of countervailing power was applied initially to the study of U.S. capitalism, where it referred to the creation of an interest organization by the people subjected to the possessors of original private economic power. An example is the formation of labor unions in the steel industry. A countervailing power is seen as performing a regulatory function. Different countervailing powers have been promoted by government in the postwar period (Galbraith 1952).

The concept developed and refers now to the presence of organized interests that represent opposing interests in a policymaking setting. The interest organizations or social movements must be strong enough in terms of organizational resources in order to counteract the power of the interest organizations that are by tradition influential in the policymaking sector (McFarland 1987b).

Empirical studies have confirmed the presence of countervailing powers in agricultural policy. They are organizations that represent consumers, environmentalists and taxpayers, that is, the consumers' movement, social movements, and labor market organizations. In Sweden the countervailing powers have been incorporated into policymaking, which has the character of a corporatist exchange.

Another research question involves the role of the countervailing powers in policy subgovernments. Scholars of public policy state that government has become sectorized into isolated policymaking subgovernments. "Entrophy of interest" or the "negotiations among the various policy-area elites in a massive system of logrolling" is said to be the result (McFarland 1987b, 10). The issue is whether the entrance of new interest organizations into a particular policy sector has cast new light on the making of policy. Research has shown that policy iron triangles change into issue networks that bring about "new forces that create strains, and change previous behavior of subgovernments and other government structures" (McFarland 1987b, 8). Studies of different countries have shown that this is the case in agricultural policy.

The final question is whether change in policy is probable when countervailing powers enter a policy sector. It has been shown in several studies that the presence of countervailing powers often changes the policy track from low to high (McFarland 1987a). It occurs since an increase in the number and type of participants adds new issues to the policy process, which enlarges the arena of discussion. The policy is transformed through *expansion* which differs from policy *narrowing*, as it does not impose established categories for classifying events and relationships (Mather and Yngvesson 1980—81). It is not surprising that agricultural policy became more consumer-oriented and is becoming more environmental, due to the role played by interest organizations that represent other groups than farmers.

When policies change from a low to a high track, conflict over the extent of the activity of the state within subgovernments is frequently a very crucial issue (McFarland 1987a). Whether government in general has become less dependent on organized agriculture as a result of the presence of a countervailing power is an important topic for research. A study of the German Farmers' Union concludes that it has happened in West Germany (Andrlik 1981). Yet it seems necessary to study other ties between government and organized agriculture before offering a conclusive answer. The ties include the effect of the rural myth on politicians and government officials, the frequency of representatives of agriculture in government and the consequences of this representation, the impact of the world market for agricultural products on the domestic situation of farmers as well as the effect of new concerns on agricultural policymaking.

2

Swedish Agricultural Politics: Acceptance of Exceptionalism

ORIGIN OF AGRARIAN POLITICAL POWER

Sweden was a nation of peasants long after most other modern European nations. Industrialism came comparatively late, with an intense stage of development in the late 1800s. Urbanization did not immediately accompany industrial development, and depopulation of the countryside occurred much later. This helps explain why Swedish peasants and people in the countryside retain a strong political position in the late 1900s. There are other important explanations as well. The historical roots of agrarian political power must be traced to the independent role that peasants played in Swedish politics. Three circumstances are important here: the fact that feudalism never dominated, the frequent coalition between the king and peasants against the nobility, and the representation of farmers in the Estate of Peasants (*Bondeståndet*) of the Parliament (*Riksdagen*).

The parliamentary estates were gradually created, fully present by the 1500s, given full legality in 1617, and abolished in 1865. The political role played by the Estate of Peasants varied between the dates. It was not very influential in the 1600s. In the Age of Freedom (1718—72), peasants were completely free personally and were more influential politically than in other European countries. The Riksdag was the most powerful political body in the period, and peasants channeled their influence through their Estate.

The Estate of Peasants cannot be considered as important politically as the other three Estates (Nobles, Clergy, and Burgers). This holds true though most of the population was somehow involved with agriculture:

more than 90 percent of the population lived in the countryside, over 75 percent was engaged in agriculture (Carlsson 1975, 18). The peasants were able to exercise considerable influence on certain political proposals and were successful in forcing through pieces of legislation dealing with land taxation and trade.

A key to peasant influence in the Riksdag was their organization. They concentrated on a few issues of economic origin that usually were very important for the peasantry itself. Also, they were able to articulate and gain support from the different estates—in particular the Clergy and Nobles—for proposals dealing with agriculture. A few Swedish peasant politicians attained a national stature not found in any other European country (Bäck 1984).

It may be concluded that political motivation and skill along with the ability to form coalitions explain the political strength of the peasantry in the Age of Freedom. Political consensus as a Swedish policy style can be said to have originated in this era, and the role played by the Peasant Estate is an important factor here. It is also interesting to note that political consciousness among peasants was enhanced in the Age of Freedom (Bäck 1984, 306—12).

A final reason for the influence exercised by the peasants in the period was their ability to use tradition as a political argument against change and for legislation favorable to the farming interest. As discussed in later chapters, reference to the Swedish rural heritage is still an important argument in politics.

The power of the peasants increased while the influence exercised by the nobility decreased in the coming years. The main political issue that engaged the peasants was the question of equality. An act that gave the peasants more rights and privileges was passed in 1789. The economic and political equality of the peasants increased in the 1800s. The negotiations on a new Swedish Constitution, ratified in 1809, illustrate the influence of the Estate of Peasants. All estates were required to accept the constitution before its adoption. The Peasant Estate opposed some of the constitutional proposals as they preserved certain privileges that discriminated against the peasant class. The other estates were forced to compromise on these issues (Stjernqvist 1975).

A bicameral parliament replaced the four estates in 1866. Election to the Upper Chamber (*Första kammaren,* FK) was indirect and guaranteed the domination of the wealthy minority. Members of the Lower Chamber (*Andra kammaren*, AK) were chosen in direct elections. Due to the principles that guided the composition of the electoral districts, the cities were to be more favorably represented in the Lower Chamber. It was a conscious choice to keep the interests of the countryside from dominating completely.

Members representing the countryside and the peasants composed initially between 40 and 50 percent of the Lower Chamber, as illustrated in Table 2.1. The farming sector was never paramount in the Upper Chamber as shown in Table 2.2, for higher civil servants and the upper class in general tended to dominate. Those members whose occupation was farming belonged to the upper class. By contrast, the farmer members of the Lower Chamber were yeomen.

In many ways, the interests that were represented in the Estates of Burgers and Peasants were still represented in the Lower Chamber. The peasant class was for many years the decisive electoral group, though the principle of apportionment gave the cities a stronger basis for representation. Their numerical power was challenged in the early 1890s when the cities began to grow. A parliamentary act of 1894 checked the influence of the cities in the Riksdag by establishing the number of seats in the Lower Chamber to 230, of which 80 would represent the cities and 150 the countryside. The act was called "the clipping of the wings of the cities" (*städernas vingklippning*) (Herlitz 1936).

Political parties became important after the 1866 parliamentary reform. Most influential in the Lower Chamber was the Farmers' Party (*Lantmannapartiet*), founded in 1867 and supported mostly by farmers. The party formulated a program that represented the interests of the agricultural class and developed an effective organization. Its domination in the Lower Chamber for 20 years can be explained by the inability of the other parties to organize themselves successfully. No crucial decision could be passed in the Riksdag against a united farmers' opposition (Carlsson 1962).

Again, organization of interest by the Swedish farmers led to political results in their favor. Though the party did not represent all categories of farmers, it was an important parliamentary instrument for the articulation of the interests of those farmers represented in the Lower Chamber. One of its ideological cornerstones was the conception of farmers as the backbone of Swedish society. The conception was based on the position of farmers as a responsible and independent social group that owned land. The party reasoned that farmers deserved a special position in politics. The line of reasoning was frequently successful. Similar political parties did not exist in the other Scandinavian countries at this time (Carlsson 1953, 1964).

Controversy over custom duties split the Farmers' Party in 1888. The supporters who favored protectionism joined the New Farmers' Party (*Nya Lantmannapartiet*), while those favoring free trade stayed with the Old Farmers' Party (*Lantmannapartiet*). The free trade issue is considered still as one of the most serious controversies in Swedish political history. The

Table 2.1
MPs in the Lower Chamber of the Riksdag with Farming as Occupation, 1867—1968, after Political Party and in Percent

Political Parties[1]	1867	1876	1885	1897	1906	1912	1922	1933	1937	1949	1961	1968[2]
Conservative	*	*	*	*	*	59	61	53	50	39	39	32
Liberal	*	*	*	*	*	59	68	54	46	25	5	8
Social Democratic	*	*	*	*	*	8	12	14	15	13	7	2
Farmer/Center[3]	*	*	*	*	*	*	95	94	89	97	71	56
Communist	*	*	*	*	*	*	14	13	0	0	0	+
Other[4]							0		14	*	*	*
All Members	52	58	51	52	47	45	43	41	37	29	21	16

Source: Adapted from Sköld and Halvarson 1966, 391, 414ff., 440ff.; Holmberg 1974, 273f.

* Not in existence or not represented in the Riksdag.

+ Data not available.

1. Some of the political parties have changed their names over the years. The ones used designate their ideological position and approximate their current names.
2. Data for the year 1968 refers to the category of farmers, including landowning and tenant farmers.
3. The parties included in the category "farmer" include the Farmers' Union and the Agrarian Party; the latter changed its name to the Center Party in 1957.
4. Other refers to political parties without much parliamentary representation. They include the Left Socialists and Socialist Party.

Table 2.2
MPs in the Upper Chamber of the Riksdag with Farming as Occupation, 1870–1962, after Political Party and in Percent

Political Parties[1]	1870	1890	1910	1937	1961
Conservative	*	*	*	49	37
Liberal	*	*	*	19	27
Social Democratic	*	*	*	2	9
Farmer/Center[2]	*	*	*	82	80
Communist	*	*	*	*	0
Other[3]	*	*	*	0	*
All Members	37	34	28	29	26

Source: Adapted from Sköld and Halvarson 1966, 401, 460ff.
* Not in existence or not represented in the Upper Chamber.
1. Some of the political parties have changed their names over the years. The ones used designate their ideological position and approximate their current names.
2. The parties included in the category "farmer" include the Farmers' Union and the Agrarian Party; the latter changed its name to the Center Party in 1957.
3. Other refers to political parties without much parliamentary representation. They include the Left Socialists and Socialist Party.

interests of producers and consumers clashed in the political debate that dealt more with economic interests than ideology.

While Members of Parliament from the cities favored free trade, MPs from the countryside were evenly divided on the issue. Political coalitions that were unthinkable in the past for ideological differences became possible. Estate owners found that they agreed with certain groups of peasants on the issue of duties. The sharp distinction between the "gentlemen farmers" (*herrar*) of the Upper Chamber and the "peasants" of the Lower Chamber (*bönder*) gradually lost its significance. Yet disagreement among different groups of peasants was apparent.

Debate on free trade dominated the political agenda from the mid-1880s to the early 1890s. The Upper Chamber supported protectionism while the majority of the Lower Chamber supported free trade. Certain duties on imported grain for example were enacted, but the issue was never decided completely. The Farmers' Party was reunited in 1894, though their disagreement over protectionism was not resolved. Rather the party decided to concentrate its efforts on representing the interests of farmers, and it played down the question of free trade (Carlsson 1953). It was probably a wise decision since farmers as a class were being challenged by newer ideologies and organized interests. Their parliamentary influence began to wane as a greater number of citizens was granted suffrage.

CHANGE IN AGRICULTURE AND POLITICS

Politics in Sweden changed greatly toward the turn of the century. The current party system began to crystallize; other political parties gained parliamentary strength as MPs started to represent them in the Riksdag; the lines of political and ideological conflict that characterize modern politics were gradually formulated. Initially the Upper and Lower Chambers of the Riksdag included different political parties. The ones that developed in the Lower Chamber are most relevant here. Unlike other political parties of the time, they had a popular basis for their standpoints.

The most influential party of the 1900s, the Social Democratic Party (*Socialdemokratiska Arbetarpartiet*), was formed in 1889. Along with the Agrarian Party (*Bondeförbundet*) from 1913, the party would come to play an important role in agricultural policymaking. A small People's Party (*Folkpartiet*) attempted unsuccessfully to represent both the interests of the city and the countryside and later joined the urban-oriented Liberal Party (*Liberala Samlingspartiet*). The Farmers' Party and the National Progressive Party (*Nationella Framstegspartiet*) from 1906 formed a moderate conservative party called "*Lantmanna- och borgarpartiet*" in 1912. Other parties were established as well.

Most of the political developments of the period were connected with industrialization. The debate on universal suffrage gained in intensity as it became obvious that many groups in society—including many peasants, rural poor, and new urban groups—did not have the right to vote. Reform was gradual. A number of disenfranchised industrial workers were granted suffrage in 1907. They used their right to vote in the 1911 election to attain a position of influence that was comparable to the one once held by the peasants. Proportional representation was also introduced for the Lower Chamber, and the Upper Chamber became more democratic. Suffrage was universal in 1921.

Industrialization led to the crystallization of an important line of conflict in Swedish politics, city versus countryside. The conflict affected all levels of Swedish society. Contrasting opinions were held by the supporters of the countryside and by the socialists and liberals. Farmers had various reactions to industrialization and the growing strength of the working class. They became politically more conservative in the 1890s. Possibly they believed that an ideological shift was a necessary reaction to the early social democratic opinion of farmers as an antiquated social class.

Other changes occurred which are important for an understanding of the situation of farmers at the turn of the century. Emigration to the United States affected the Swedish mentality greatly. Over one million Swedes emigrated in the 1800s. The largest number left in the 1880s, an event explained by the agricultural crisis of the period. The aforementioned political debate on free trade and the enactment of certain agricultural duties in 1888 was one early political reaction to emigration. Emigration began in the 1840s and virtually ended by 1930 (Alestalo and Kuhnle 1984).

Initially the government was not really alarmed by the development: the countryside was overpopulated, emigration was seen as a solution to the problem, and it was generally considered that those who chose to emigrate were dispensible. A change in attitude occurred.

Part of the explanation of the change lies in the realization that the groups that emigrated were important for national well-being. A number of men who were of draft age and young sons and daughters who lived and worked on farms emigrated. In an era of growing nationalism and patriotism, it was felt that "a flood of Sweden's best blood" poured into other countries (Carlsson 1966, 38). Emigration was the social occurrance that received the most attention in the first decades of the 1900s.

People living in the countryside sought work in industry. Historians tend to view the transformation as a cultural shock that shook Sweden harder than any other West or Northern European nation. The number of people engaged in agriculture plummeted from 73 percent of the population in 1870 to 44 percent in 1920 (Carlsson 1956, 309). Large numbers of farmers became wage earners (Carlsson 1956, 1966; Elmér 1975).

Depopulation of the Swedish countryside occurred rather late in comparison to other countries of similar nature. Urbanization and industrialization were not simultaneous developments, which is explained by the formation of smaller rural towns in the period. Demographic change was uneven: the northern areas of Sweden and small farmers were most affected.

The movement of people into the cities is striking. In 1866, 12 percent of the Swedish population lived in cities. By 1945, 42 percent were city-dwellers. Twenty years later over 70 percent lived in cities and densely populated areas (*tätorter*), a type of residence not very common in the 1860s. By the 1980s, less than 20 percent of the population lived in the countryside (Carlsson 1966, 3ff.; SCB 1946, Table 4; SCB 1988, Table 22). "Almost a whole nation moved from farming villages, crofts and cottages to highrises, tenements, terrace houses and suburban homes" (Carlsson 1968c, 9, my translation).

Farmers decreased in number as well as importance, while the working class and the urban areas grew in strength. It meant that cities began to demand and receive more political attention. Political parties catering to more urban needs were created and gained in strength. Members of Parliament resided in and represented urban areas in growing numbers. Table 2.3 il-

Table 2.3
MPs in the Lower Chamber of the Riksdag Residing in the Countryside, 1912—1961, after Political Party, in Percent

Political Party[1]	1912	1922	1933	1937	1949	1961
Conservative	67	65	57	59	57	39
Agrarian/Center[2]	*	100	94	94	90	85
Liberal	77	73	67	48	42	20
Social Democratic	52	40	45	52	46	36
Communist	*	71	63	40	33	0
Other[3]	*	50	*	50	*	*
All members	67	59	59	60	51	40

Source: Adapted from Sköld and Halvarson 1966, 412, 438.
* Not in existence or not represented in the Lower Chamber.
1. The political parties included in this table have not always had the same names. The ones used designate their ideological position and approximate their current names.
2. The Agrarian Party was established in 1913. It changed its name to the Center Party in 1957. The statistics include parliamentary representation of the smaller Farmers' Union.
3. Other includes the Left Socialist Party and Socialist Party.

Table 2.4
MPs of the Upper Chamber of the Riksdag Residing in the Countryside, 1921—1961, after Political Party and in Percent

Political Party[1]	1921	1929	1937	1961
Conservative	49	41	60	47
Agrarian/Center[2]	84	88	77	90
Liberal	44	42	56	24
Social Democratic	14	23	20	39
Communist	*	0	*	0
Other[3]	25	*	0	0
All Members	40	41	44	43

Source: Adapted from Sköld and Halvarson 1966, 425, 460.
* Not in existence or not represented in the Upper Chamber.
1. The political parties included in the table have not always had the same names. The ones used designate their ideological orientation and approximate their current names.
2. The Agrarian Party was established in 1913. It changed its name to the Center Party in 1957. The statistics also include parliamentary representation of the smaller Farmers' Union.
3. Other parties include the Left Socialists and Socialist Party.

lustrates the change by comparing the number of MPs in the Lower Chamber from the cities and the countryside. Table 2.4 offers similar information for the Upper Chamber. Not only did the agricultural sector lose political power and a sound economic basis for growth, it also began to lag behind other sectors, such as communication and transportation, in technology.

Without exaggeration it can be said that industrialization and urbanization created a new nation. Yet the transformation was not complete. To this day many people view industrialization and urbanization as negative developments. The opinion is often expressed as a distrust and dislike of a centralized state and a romantic view of the countryside. Both expressions have long historical roots.

"Anti-bureaucratic" sentiments can be traced to the rural individualism that once characterized Swedish farmers. An extreme example comes from the 1500s. The rather inferior infrastructure of the time and the presence of massive forests made it easy for farmers to block roads with large logs taken from the forests to prohibit the military from reaching its destination. Also, they could refuse to pay rent fees and taxes by simply dismantling their log cabins and rebuilding them at another place (Lönnroth 1989).

The social mobility of the 1800s modified their individualism and brought the peasants in greater contact with the state. Developments as industrialization and urbanization demanded that government assume a new, different, and more active role. Due to their ingrained distrust of civil servants and the "*overhet*" (see Kelman 1981, 118ff. for a discussion of the concept), peasants initially opposed the more active role of government. Compulsory military service, national defense, and taxation were important issues. The policy of the Farmers' Party was thrift, caution, and debureaucratization (Kihlberg 1922, 499). This view was modified somewhat in the following years, though the last issue is of current relevance.

In 1914 the farmers demonstrated that they were patriots when they (roughly 31,000 in number) met in Stockholm with a petition signed by 40,000 farmers. The "Farmers' March" (*Bondetåget*) was a reaction to a long and serious political debate on defense policy and expenses. With the mass demonstration, the farmers wanted to declare directly to the Swedish king—but not to the government—that they were willing to bear increased defense costs. The form of political expression was unusual for the time. The demonstrations held by the working class on May Day were the example followed by the farmers.

Their appeal perpetuated the notion of a band of trust between the farmers and the Swedish king as well as the conception of the farmers as the one social group that represents the Swedish people. Not only did the event touch off an awaiting political crisis, it forced the government to submit its resignation and call a new election.

It showed that Sweden was really two nations: one of the declining countryside that used organization and romanticism to influence politics, the other of the burgeoning industrial class that also used organization to gain influence but had the potential for growth on its side (Strandberg 1968; Schiller 1975; Stenkvist 1987). The Farmers' March showed that organization of farmers was again an advantageous strategy.

A new era of peasant romanticism began in the last decades of the nineteenth century. As peasant life was threatened by change, its traditions and activities were increasingly praised in poems, paintings, education, genealogical research, and museums. The Nordic Museum in Stockholm was build in the 1870s and houses an impressive collection of Nordic cultural history, including a special wing on the people of the countryside (*allmogen*). It is a home for memories, in particular those from the life of the Swedish people (*Nordisk familjebok* 1913).

Its annex is the open-air museum Skansen in Stockholm that dates from the 1890s and aims to show new generations the life of their forefathers. Skansen is world-renowned as the first museum of its kind and as a prototype for ones in other countries. It had over 770,000 visitors in 1910 and roughly 1,700,000 in 1988.

Though depopulation of the countryside continues, Skansen and the Nordic Museum instill urbanized Swedes with a feeling for the past. Both interest people in their homesteads, protection of archeological remains as well as traditional arts and crafts. In typical Swedish fashion, the interest has also taken organizational form. One of Sweden's oldest popular movements, the local culture society (*hembygdsföreningen*), formed a national organization in 1916. By 1920, there were 200 local societies nationwide. The current organization, the National Association for the Preservation of Local Culture and Nature (*Riksförbundet för hembygdsvård*), represents 25 regional units and about 2,000 local ones. Other organizations that romanticize the countryside would be created in following years.

As an art form that expresses sentiment for the countryside, national romanticism had a number of important repercussions. Its characteristic motif is the homestead. Along with other romantic reactions, it developed as the countryside underwent transformation. It was "love of that which was destined to disappear, 'the red cottages and manorial estates,' the agrarian society and the wilderness, folklore and popular tunes" (Andreae 1968, 197, my translation) Two well-known Swedish artists, Anders Zorn and Carl Larsson, were national romantics. Their art affected a number of famous Swedish writers, and Larsson's paintings directly influenced urban interior decoration. They are a few ways in which the romantic sentiment for the countryside have been passed on through the generations.

The countryside also developed into a retreat for urbanites in the period. Love of the land became a symbol for Swedishness and vice versa. City people looked forward to a trip into the countryside. National romanticism as an art form most assuredly furthered this development. It is noteworthy that the class consciousness so evident in the cities in this period was not applied to the countryside. Myths were created as the atypical Swedish province of Dalarna—with its small number of farm worker residents (and thus low level of class conflict) with their picturesque provincial dress—came to symbolize the Swedish farmland and peasantry. The tourist industry capitalized on the urban association of the countryside with peace, quiet, and goodness (Frykman and Löfgren 1987).

The countryside, nature, vacation, and summer symbolize the good life to the sunworshipping and overworked urban Swede. An important conclusion here is the imbalance of sentiments. While urbanites evaluate the countryside in positive terms, rural inhabitants view the cities with mistrust and animosity. It uncovers an underlying explanation for why Swedes react with a feeling of guilt when unfortunate developments occur in agriculture and in the countryside. An anti-rural myth exists, but it has not exercised as strong an influence. The political and interest organizations of farmers have played upon the feeling of collective guilt to attain their political goals. Similar reactions occurred in other countries as well.

INITIAL INVOLVEMENT OF STATE IN AGRICULTURE

Agricultural policy can be traced far back in history. Grain policies were common in the Middle Ages, but policymaking occurred usually on the local rather than national level. Formulation of a national rather than local policy demanded that the cities cooperate with each other and reach a compromise. Often compromise could not be reached readily.

The situation in Sweden was different. The Swedish national government has almost always played a strong role in agricultural policy. The explanation is the relatively late development of cities. The government was involved in the collection, storage, and provision of grain in emergencies at a rather early date. Another reason for the strong role of the state is the comparatively late development of the retail sale of grain. In many other European countries, grain trade was an important business. In Sweden, bartering continued to dominate as the primary means of trade in the 1800s (Åmark 1915).

The involvement of the Swedish government in agriculture increased.

From the 1700s, the county agricultural societies (*hushållningssällskapen*) gave advice to farmers and took part in the establishment of institutions for agriculture. Public schools of agriculture date from the 1830s; schools that educated farmers were established in the 1860s. The Riksdag legislated on taxation of farm property and land use, and a special parliamentary committee on economic problems of agriculture was formed in 1867. Yet when compared to the role that the state would come to play, its activity in agriculture in these early years was limited (Gellerman 1958, 10f.).

More active involvement occurred at the turn of the century with the formation of the agricultural policy sector. As agriculture became threatened in different ways, it was generally felt that more regulation was necessary. The government offered protection by creating an agricultural policy sector.

The National Board of Agriculture (*Lantbruksstyrelsen*, LBS) was created in 1890 to assume many of the tasks performed by the Royal Academy of Agriculture (*Lantbruksakademien*) from 1811. In 1900, the Ministry of Agriculture (*Jordbruksdepartementet*, Jo) was established in order to relieve another ministry of its workload in this area and to accommodate the desire that was expressed in the Riksdag and by farmers on this issue. Quite a number of MPs whose constituencies were urban opposed this decision (Gellerman 1958, 51). The Parliamentary Standing Committee on Agriculture (*Jordbruksutskottet*, JoU) replaced the temporary one in 1909. Other government bodies that were involved in agriculture and forestry were established in these years.

One of the serious problems facing the embryonic agricultural sector was emigration. The agricultural county societies submitted reports on its effects on the agricultural sector; a royal commission was appointed to investigate the problem, and two royal commissions (one appointed in 1899, the other in 1914) discussed what the government could do to halt emigration and depopulation of the countryside.

The author of the report by the royal commission on emigration was successful in influencing the public debate when he romanticized on the love that the farmer has for the land (Wohlin 1919). The other commission suggested that government loans for the purchase of land might help curb emigration. This was called the Homestead Movement (*Egnahemsrörelsen*). The loans did not solve the problem.

The movement itself was enveloped in the peasant romanticism of its time that viewed the landowning farmer as possessing certain valuable and special personality traits and considered farm life as character-building. It was a romanticism that would develop into the rural myth. What is particularly interesting about the homestead committees is that they emphasized more the social and demographic aspects of the problem than the strict economic and agricultural ones (Wohlin 1919, 109—34; Elmér 1975, 44ff.; Gellerman 1958, 358).

POLITICS OF CRISIS

The two world wars affected Sweden in many ways. Both wars caused a number of serious problem for the Swedish state and people. One of them was national food supply. The situation was particularly acute in World War I (1914—18), as Sweden was forced to become almost completely self-sufficient after years of increased dependency on imports. Plans for food supply and food distribution were not made at the outbreak of the war, due to misjudgments regarding the probable length of the war. A number of bad harvests amplified the problem.

When compared to how a similar situation was dealt with in World War II (1939—45), another reason for the food crisis of the first war was the weak organization of farmers in both a union and a cooperative movement. Not only were there deficiencies in the system for the distribution of food. It was difficult to convince farmers to obey the regulations and behave in a publicly responsible way (Gellerman 1958, 353; Carlsson 1968c; Heckscher 1951, 231—37).

It is not an easy task to summarize briefly Swedish agricultural policy in World War I. Change in government occurred, and this affected agricultural policy. Producers and consumers disagreed over solutions to the problem of food supply. The conflict increased as the situation worsened and the war continued. Agricultural policymaking of the period reflects the problems and also a balance between the conflicting interests. Four general guidelines were given different emphasis, depending on the political situation. They included: avoidance as much as possible of state involvement that would disturb the farming operation; government measures that would ensure national supply of staple foods; checks on speculation and inflation; and economic measures to stimulate farming (Gellerman 1958).

Representatives of the consumers favored government action and regulation. Representatives of the producers were opposed. Their opposition was successfully channeled in the later years of the war through two political parties, the Agrarian Party and the Farmers' Union (*Jordbrukarnas Riksförbund*), that explicitly and unreservedly represented the interests of farmers. A farmer and spokesman of this group was appointed as minister of agriculture in the liberal/social democratic coalition government (1917—20).

The agricultural policy of World War I was decided after harsh political debate. Regulation became more coercive as the crisis grew and involved price indexing, rationing, and bans on exports (Hedlund and Lundahl 1985a). Adamant social democratic articulation of the interests of the consumers eased the way for the early efforts of regulation. The social democrats supported war socialization also. The Commission on Food (*Livsmedelskommissionen*) was appointed in 1915. Later the social democrats toned down the interests of consumers, since there were a number of farmers within the party, and criticized instead the middlemen for the high food prices (Hedlund and Lindahl 1985a, 237, 259, 266f., 352).

A number of lessons were learned from the war experience. Self-sufficiency in agriculture began to be considered as an essential part of the Swedish doctrine of neutrality. Interest organizations could play a role in the coordination of public activities in times of crisis. The importance of political compromise, consensus, and stability for successful crisis planning and policy became apparent. All the lessons implied a new role for Swedish farmers, and they benefited the farmers economically as well as politically.

Self-sufficiency in food production is a part of Swedish national defense. Previous to World War I, war preparedness was strictly viewed in military terms. Experience from the war taught the governments of many countries that preparedness must be "total" and not only include military but also civil, psychological, and economic defense.

In 1928, the National Committee for Economic Defense (*Rikskommissionen för ekonomisk försvarsberedskap*) was appointed and given the task of preparing Sweden for crisis. One issue was national food supply crisis planning. Since its authority and resources were limited, the commission was not successful in formulating a general policy of or plan for war preparedness covering the vulnerable aspects of Swedish government and society (Månsson 1976).

War preparedness was to become a main principle of postwar Swedish agricultural policy. Regulatory measures were employed to ensure self-sufficiency. A number of measures were introduced in the early 1930s, many of which were a reaction to the world depression that began in 1929. As is often the case, the state attempted to regulate and later stabilize the agricultural market by creating an administrative apparatus. Many of the current public bodies in the agricultural policy sector find their origin in the institutions from the inter-war period. Similar developments occurred in other western countries as well, for example the United States (Galston 1985).

A number of new institutional bodies were given broad responsibility to regulate the import, export, and storage of agricultural commodities. Frequently interest organizations that were affected by the intervention sat on their governing boards. The government and interest organizations cooperated very closely in the inter-war years (Ovesson 1958; Gustafsson 1974; Åkerman 1937; Friberg 1973). A corporatist pattern of exchange developed. Probably the most interesting example concerns milk regulation, as discussed in Chapters 3 and 5.

Other examples of the pattern of exchange include the institutions developed to regulate the marketing of farm products. A board, the Swedish Grain Association (*Svenska Spannmålsföreningen*), was established in 1931 to regulate the grain market, which involved the government in the internal affairs of the farmers' cooperatives. The Swedish Farmers' Supply and Crop Marketing Association (*Svenska Lantmännens Riksförbund,* SLR) was reconstructed by the government in order for it to be able to participate effectively in grain regulation. The government wanted the farmers organized into an association that it could work with. The social democrats were instrumental here. The Grain Association (renamed Swedish Grain Trade Association, *Svensk Spannmålshandel*) still exists as the sole Swedish importer and exporter of grain. It purchases all Swedish wheat and rye of suitable grinding quality at a price set by the government. By 1941, it had 114 employees.

To a large extent, the inter-war years formed the agricultural policy network. Active government involvement in agriculture developed in the period. Interest organizations for farmers were created, and farmers began to view government intervention in a new light. It is important to note that the agricultural policy that developed as a reaction to the inter-war agricultural crisis came to be the point of departure for postwar policymaking. Inter-war policies were shortsighted and attempted to solve single problems as they arose. Policymaking was ad hoc and incremental. Two consequences were policy solutions that contradicted each other and unintended consequences of the policy goals (Ovesen 1958).

Policymakers were aware of the problem at the time but had difficulty solving it. To this day, attempts to solve the illogic of the situation without the help of the farmers' organizations are usually not successful. An attempt to coordinate the activities of the regulatory market boards was made when the National Agricultural Market Board (*Statens Jordbruksnämnd,* JN) was created in 1935. Further reevaluation of the agricultural policy jungle was interrupted by World War II.

Agricultural policy of the inter-war years did not only involve market regulation that was based on the principle of high protectionism. Different kinds of direct economic support were also given to individual farmers. They included subsidies, loans, income supplements, and credit guarantees (Hellström 1976). Producers—especially larger farmers—were favored by these policies and consumers, in particular those with low incomes, were disfavored. The general reason for the bias in policy is the influential position that the farmers' organizations were given when they interacted with government.

World War II meant another crisis for Sweden. The focus of agricultural policy was changed. Inter-war protectionism was no longer necessary, as there was an embargo. Income compensation became the main policy goal, and agricultural regulation aimed at increasing domestic production. As in the past, new administrative bodies were created to take on the tasks. Around thirty public bodies dealt specifically with the agricultural sector, many of which were under the authority of the central regulatory organization, the National Commission of Food (*Statens Livsmedelskommission*) from 1939. Included in a council associated with it were representatives for the farmers' associations, the Cooperative Union and Wholesale Society (*Kooperativa Förbundet,* KF), Swedish Retail Federation (*Köpmannaförbundet*), and national association of home economists (*Sveriges husmodersföreningars riksförbund*).

The authority for intervention by the state was vast. The Riksdag sanctioned price and wage stops and enacted a law on compulsory cultivation of sugar beets. Grain was confiscated by the state. Egg production was directed through controlled sale of chicken feed and state-sanctioned wholesalers. Regulatory bodies existed on the national, regional, and local levels. That the farmers' organizations cooperated in the regulation is a very important reason for its success in a situation of international agricultural crisis, bad harvests, and insufficient food supplies (Åmark 1952; Friberg 1973; Olsson 1945).

PARTY POLITICS AND FARMERS

The predicament of farmers has generally met with sympathy from the major political parties of Sweden. The importance of agricultural self-sufficiency for Sweden is one essential reason as is the effect of the rural myth on Swedish attitudes toward farmers. The creation of specific interest organizations for farmers is a third. Finally, there is one explanation that applies particularly to the Scandinavian countries. It is the role of farmers and their representatives in the politics of compromise. This section focuses on the Social Democratic and Agrarian Parties, the two political parties most influential in agricultural policymaking in Sweden.

The social democrats initially considered farming as a depressed and stagnating employment sector and farmers as an antiquated social class. Yet agriculture was never really severely challenged politically by the growing influence of the industrial class. The party did not seriously attempt to exploit the crisis situation of farmers by formulating a specific interest of workers as consumers that pressured for high quality food at low prices. Rather, social democrats at first chose to see the farm problem as one of economics and markets. Farmers were a potential ally (*see* Thullberg 1974 for a general discussion).

Social democratic thinking on the agricultural problem changed over the years. Though the socialist standpoint on nationalization of land was a hotly debated topic, sympathy developed in the 1890s for farmers owning small farms and tenant farmers. It was influenced by the debate among socialists in other countries, particularly England, Germany, and Denmark. Support for farmers' cooperatives as an acceptable form of ownership also came early. They were seen as a way to give small farmers the benefits of large-scale production. All leading spokesmen did not agree on the importance of the smaller farmers for the furtherance of social democracy.

The party program of 1911 marks a drastic change from past thinking and the first active social democratic policy on agriculture. Farmers were considered as workers who owned their own means of production. It recognized and gave legitimacy to small farmers, though farm nationalization was still part of the party platform. The 1920 program went further and established that small farm owners and workers belonged to the same class. The reasons for the ideological revision are the opinions of party sympathizers, emigration, peasant romanticism, foreign influence, and structural changes in agricultural production (which favored small farmers) (Tingsten 1941; Odhner 1988).

Vote maximization played a role in party thinking as well. The 1909 voting rights' reform meant that many more farmers would vote in the 1911 election. All political parties attempted to formulate a more cogent policy on agriculture. Farmers continued to be an important political force. Political measures to help the agricultural sector were exchanged with measures to help the working class. The first trade-off occurred in 1919, when the social democrats agreed to support a duty on agricultural commodities in turn for non-socialist support for the eight hour workday.

The Social Democratic Party had supported a policy of free trade. It was thought that free trade would lead to low food prices, the maturation of capitalism, and further the development of Swedish socialism. Gradually, the issue of free trade led to controversy within the party. Leading social democrats were for and against protectionism. A less influential group on the far left wanted to exasperate the conflict between consumers and producers. The party decided to support protectionism and develop the corporatist exchange.

Social democratic acceptance of protectionist measures for agriculture was made public in 1932, when the party created a minority government. It decided to retain the regulations that were enacted by the former governments, and one of the leading farmers and promotors of protectionism within the party, Per Edvin Sköld, was appointed minister of agriculture. Together with Ernst Wigforss, the other leading supporter of protectionism, he formulated a proposal for an excise duty on margarine as an enticement for either the Agrarian Party or the Prohibitionist Party (*Frisinnade Folkpartiet*) to engage in a political exchange. In turn, the social democrats wanted parliamentary support for their unemployment policy. They made a historic compromise with the agrarians in 1933. An earlier attempt to reach an agreement was made in 1929 when they negotiated unsuccessfully with the prohibitionists.

Social democratic understanding of the farm problem is not astonishing. Ideologically it was decided that farmers and industrial workers played the same role in the capitalist system. As argued convincingly by Sköld in an article in the party's ideological journal (*Tiden*) in 1926, class consciousness was considered as necessary for farmers as workers. He reasoned further that organization of the interests of farmers in both union and producer associations should be promoted for ideological and economic reasons, and it was not an issue for party politics. He sought support for this viewpoint from the other parties in the Riksdag. Another leading farmer who was a member of the governing board of the party argued that higher food prices might be necessary to improve the economic situation of the farmers and that workers should be taught to understand this.

The farmers who held influential party posts could influence social democratic ideology in significant ways. It is not surprising when consideration is given to their numerical weight: between 40 and 50 percent of all social democratic MPs in the Lower Chamber resided in the countryside, and a good number had farming as their occupation, as illustrated in Tables 2.1 and 2.3. The party's agricultural experts were in agreement with those farmers who held similar posts in the other parties.

Discussions on agriculture inside the party generally dealt with questions of economic and social policy. Yet the standpoints taken by the party were not always consistent. While small farmers were promoted for reasons that dealt with social policy, larger farms were seen as the most efficient and effective from the point of view of industrial policy.

Swedish agricultural policy during most of the 1900s has, in reality, been a social policy for farmers. That agriculture came to be excepted did not create controversy. In fact, it can be argued that the agricultural policymakers within the Social Democratic Party were successful in uniting the interests of consumers and farmers. One example is recruitment of farm laborers to the Swedish Union of Agricultural Workers (*Svenska Lantarbetareförbundet*), which did not lead to criticism by either the Social Democratic or Agrarian Parties. The union joined the Confederation of Swedish Trade Unions (*Landsorganisationen i Sverige,* LO) in the 1930s. Also, the policy of agricultural protectionism became an aspect of the party's program for a planned economy (*planhushållningen*), which sanctions state control of the economy (Holmström 1983; Thullberg 1974; Lewin 1967).

Social democratic ideology toward agriculture changed greatly over a period of 50 years. First it was marxist. It then became liberal in the early years of this century. Later it became protectionist, a policy standpoint dating from the late 1920s and greatly affecting Swedish national policymaking. Now it is changing toward an environmental and free-trade orientation. Deregulation is emphasized.

Peasant romanticism played a part in social democratic thinking, though never as strongly as for the Agrarian Party or early conservative parties. The party's interest in the homestead movement was based in large part on a romantic conception of landownership. Its support of small farmers and the family farm is tied to a romantic conception of farming as a way of life or, as denoted in Chapter 1, the rural myth. Influential farmers within the party promoted romanticism.

Romantic notions can also be found within another workers' party of the period, the Swedish Communist Party (*Sveriges kommunistiska parti*). Along with workers, small farmers were initially seen as belonging to the exploited class. Its 1918 congress went so far as to state that individuals should be brought up to love the land. Later, the communists encouraged cooperation between unionized farmers and workers.

It is difficult to assess whether ideology, romanticism, or political tactics is the most important explanation of the agricultural policy that was adopted by the social democrats. Political scientists often view vote maximization as an important explanation of the behavior of political parties. Undoubtedly the policy stands taken on the agricultural issue were also an electoral tactic for the social democrats. The agrarians and social democrats competed with one another for the votes of small farmers as well as farm workers as early as the 1920s (Simonson 1985; Hellström 1964, 86ff.).

Vote maximization may be conceived as a political trade-off that occurs within a party. Rather than reaching a compromise with another party in order to gain a parliamentary majority, the party undergoes an internal compromise and, thereby, hopes to gain enough votes in order to be able to implement its policies.

The Swedish Social Democratic Party underwent this change. It began as a working class party struggling for the proletarian cause and then became a people's party representing the national interest. It created the People's Home (*folkhemmet*) (Esping-Andersen 1985). The transformation occurred in the 1930s, a time when 68 percent of the Swedish population lived in the countryside, 21 percent had farming as an occupation, 58 percent were blue-collar workers, and 41 percent of the electorate voted social democratic (SCB 1935, Table 9; Birgersson et al. 1984, Tables 1C, 6C).

The interest that was shown by the Agrarian Party in agriculture is, at least up to the 1950s, much easier to understand. The party was created as a reaction to industrialism (the new economic situation of the early 1900s), the liberal orientation of the conservatives, and the inability of the existing parties to safeguard the interest of farmers. From the beginning, the Agrarian Party and the Farmers' Union (*Jordbrukarnas Riksförbund*), a more conservative wing of the Agrarian Party that left in 1915, sought to represent farmers as both a political party and an interest organization, an idea they copied from the workers' movement.

After the 1917 election, both parties gained enough votes for representation in the Lower Chamber of the Riksdag. The next time a new party (the Environmental Party/the Greens, *Miljöpartiet/De Gröna*) entered the Riksdag was in 1988. The agrarians were approached before the election by the conservatives who invited them to campaign under a united front. While the Agrarian Party declined, the Farmers' Union accepted. The parties began to build up their party organizations at this time. A youth organization and a women's organization were formed.

After years of dispute and discussion, the two parties agreed to amalgamate in 1921. The new party retained the name the Agrarian Party. Political organization of the agricultural interest was successful, as seen in Table 2.5. Small farmers were mobilized and organized into politics at an uncomparable rate. Their electoral participation gave the party enough votes to lead to parliamentary representation, a circumstance that encouraged different political parties in the 1920s to suggest cooperation in either electoral campaigns or in the building of governments.

Though it began to give attention to more general issues regarding the countryside in the early 1930s, the Agrarian Party was still considered to be a farmers' party. Yet it never succeeded in mobilizing the entire agricultural population. Support came mainly from small farmers residing in the southern parts of the country. Neither farm laborers nor large estate farmers showed great sympathy. The party was closely connected with the National Farmers' Union (*Riksförbundet Landsbygdens Folk,* RLF), and most of the leaders of the RLF were members of it.

Table 2.5
Political Strength of the Agrarian Party, 1914—1957, in Percent

	Elections to the Riksdag													
	1914	**1917**	**1920**	**1921**	**1924**	**1928**	**1932**	**1936**	**1940**	**1944**	**1948**	**1952**	**1956**	
Votes[1]	0.2	9	14	11	11	11	14	14	12	14	12	11	9	
	Seats in the Riksdag[2]													
	1914	**1915**	**1918**	**1921**	**1922**	**1925**	**1929**	**1933**	**1937**	**1941**	**1945**	**1949**	**1953**	**1957**
Lower Chamber	0	0	6	13	9	10	12	16	16	12	15	13	11	8
Upper Chamber	0	0	0	13	12	12	11	12	15	16	14	14	17	17

Source: Adapted from Birgersson et al. 1984, Tables 1C and 2A.

1. Refers to popular elections to the Lower Chamber and to the assemblies that selected MPs to the Upper Chamber.
2. Includes seats for both the Agrarian Party and the Farmers' Union in 1918—1922.

The party stressed that it represented the countryside against the cities and stood for the role of popular movements in society against bureaucratic influences. The role of farmers and farm owners in Swedish history was emphasized in criticism of urban bureaucratization. Peasant romanticism was still an important aspect of agrarian politics as farmers were placed above other social groups.

Also, it considered itself to be a centrist party that was placed in between the parties of the left and the right. The central position has even led to a description of the party as a "joker" in all discussions in the 1940s and 1950s on cooperation among the non-socialist parties (Näsgård 1936; Urwin 1980; Petersson 1981; Torstendahl 1969). The Agrarian Party strove to broaden its basis after its participation in the historical compromise of 1933 and later in coalition governments with the social democrats.

IMPLICATIONS OF HISTORICAL COMPROMISE FOR FARMERS

The agreement reached between the social democratic minority government and the Agrarian Party in 1933 is considered to be a historical compromise for many reasons; many of which will not be touched upon here. For the farmers, it meant that their interests were to become strongly united with the government apparatus. It had a number of important consequences for the interest organizations of farmers. What is particularly interesting is that the Agrarian Party chose to cooperate with the Social Democratic Party rather than participate in a non-socialist coalition or assume a more conservative position, as had occurred in other European countries. A number of events explain this.

Small farmers began to identify themselves with workers in the agricultural crisis of 1929. The association of farmers' and workers' interests was promoted by a new generation of agrarian politicians and by the RLF, an organization that developed quickly and was patterned after a trade union. It was these groups within the Agrarian Party that encouraged agreement with the Social Democratic Party.

The compromise signified that the Agrarian Party changed its outlook. It was open to new ideas from the RLF on how to solve the farm problem. Farm income was to improve through rationalization, effectivity, specialization, and the organization of farmers into unions and cooperatives rather than through increases in food prices and custom duties, as supported by the Conservative Party (Thullberg 1972).

The political backdrop for the 1933 compromise needs to be understood. As is often the case with these kinds on political arangements, the considerations that lead to agreement are immediate problems. Yet the agreements themselves reflect more than their immediacy. A number of minority governments were formed in the 1920s. Executive power was weak, and the stronger political parties desired to gain sufficient parliamentary support for longevity in office. Though Sweden was less affected than many other countries, the depression of 1929 added to the political judgment that stability was a foremost national priority.

A sharp division of the parties into a socialist and a non-socialist bloc developed after universal suffrage was introduced. The division was particularly sharp on the issue of unemployment. The political debate in the Riksdag in 1932 was concentrated on the issue of unemployment as well as financial support to farmers. There was disagreement within the non-socialist bloc on several issues, among them protectionism, which prevented the formation of a non-socialist government. In the election of 1932, which was to a large degree a struggle over votes from the countryside, the social democrats, communists, and agrarians gained a number of parliamentary seats at the expense of the conservatives and liberals, as shown in Table 2.5. The socialist bloc was 7 seats short of a majority in the Lower Chamber of the Riksdag.

The Social Democratic Party formed a minority government after the election. Prime Minister Per Albin Hansson was a critic of weak executives and desired to broaden the strength of his own newly formed government. Also he reasoned tactically: if the party were to increase its electoral base in the countryside, it must revise its agricultural policy and support public measures to help farmers.

He spoke of the similarity between the situation of farmers and workers: "the problems of the workers are the problems of the farmers, and the problems of the farmers are also the problems of the workers" (*"arbetarnöd är bondenöd, och bondenöd är också arbetarnöd"*) (from his May Day speech as quoted in Nyman 1944, 26, my translation). Other leading social democrats were in agreement. At the same time, social democratic agricultural policy was undergoing change, as was that of the Agrarian Party. These are the less immediate explanations to the historical agreement (Hellström 1976, 484—503).

The 1933 agreement is frequently referred to in discussions of the political culture of Sweden. It is said to show that political parties and movements which formerly opposed each other can reach a compromise that is mutually favorable (Tingsten 1941, 361). The agreement was called horse-trading (*kohandel*) in the public debate. The social democrats declared openly that they favored agricultural regulation and protectionism and promised to support farmers. In turn, the Agrarian Party promised to support parliamentary proposals from the government on unemployment policy (Hellström 1976, 514—47). Afterward, the new Liberal Party from 1934 (*Folkpartiet*) decided to support the agreement.

Though the inter-party cooperation initially met with some criticism, it proved to be beneficial for both parties. They gained at the polls in the 1936 election, as seen in Table 2.5. The system of agricultural regulation continued to develop in a political atmosphere that did not criticize it. Regulation was said to promote both stability and reliability in production as well as a reasonable distribution of welfare.

Cooperation between the two parties continued. After the election of 1936, the parties entered a coalition government. The leaders of the Agrarian Party, Axel Pehrsson-Bramstorp, became Minister of Agriculture. Pehrsson-Bramstorp was one of the leader of the party who argued for the new party outlook on agricultural policy. He favored the unionization of farmers and their cooperation with trade unions. The coalition stayed in office until World War II, when a four-party coalition government (including the social democrats, agrarians, liberals, and conservatives) was formed.

POSTWAR SWEDISH AGRICULTURAL POLICY AND POLITICAL CONCERNS

The experiences in the war crises greatly affected agricultural policymaking in the postwar period. It became clear that deregulation of agriculture would not be possible. The political parties had, otherwise, wanted to deregulate after the wartime crisis, but the international situation prohibited this. Other countries enacted support systems for farmers, and Sweden would be forced to do the same. The importance of war preparedness also affected thinking on agricultural policy. Food policy was put in the focus.

Four important parliamentary decisions formulate current Swedish agricultural policy. The first is discussed below and the remaining ones in Chapter 4. The decisions were prepared by royal commissions, which were appointed by the government to study the subject of agriculture. The commissions typically include members who represent the different political parties, public agencies, and involved interest organizations.

A commission was appointed in 1942 by the Minister of Agriculture, who was also the leader of the Agrarian Party and who acquiesced to demands from the Swedish Agricultural Society (*Sveriges Allmänna Lantbrukssällskap*, SAL), to study the subject of agricultural policy in peacetime. The interests of agriculture were well represented on the commission: half of the members, including all the social democrats, can be so classified. The commission's reports formed the basis for the 1947 parliamentary decision on agriculture, which has been called the "Magna Charta of agricultural policy" (Swedborg 1980, 19).

What is particularly interesting about the commission's reports is that they focused on the economic aspects of agriculture and not social considerations or the standard of rural life as had been done in the past. Farming was to be understood as a business, not a way of life. The view reflected in the reports is agricultural realism rather than romanticism. Agricultural exceptionalism was to come to an end. Emphasis was put on economic efficiency through rationalization and modernization. The solution to rural unemployment was industrial and urban employment. Schemes that attempted to remedy the gap between the income of farmers and industrial workers were seen as easily in conflict with the economic considerations contained in the final report. It was argued that war preparedness was an essential aspect of agricultural policy (Thullberg 1980).

The reports written by the commission and the remiss comments on them is an example of the way different opinions on agricultural policy would be articulated in the future. Today economists support deregulation and view agriculture as a business that should be exposed to competition. Many of the experts whom the commission relied upon were economists, whose opinions are somewhat similar to the current one. Public agencies hold another view that stresses policy incrementalism, the influence of organized interest on decisionmaking, and perhaps the desire for continued organizational existence. The trade unions fit somewhere in between these views. Though the unions have come to represent the interest of the consumers, they have often found it difficult to voice adamant criticism of the agricultural situation.

The parliamentary bill presented by the Social Democratic government in 1947 did not stress the concerns of the role of agriculture in the national economy. Rather, Minister of Agriculture Per Edvin Sköld emphasized income equality as the key goal. War preparedness was given priority as well.

Farming was not to become a business to be transformed by technical advancement, and the rationalizion of farms into larger units was to be restricted. No upper guideline was formulated on agricultural production. The family farm was established as a principle, and special economic support was alloted to small farmers. The committee's proposal on giving economic support to individual farmers to further farm rationalization was not accepted by the minister. Agriculture retained its privileged position and was excepted from the general transformation that society underwent in this period (Odhner 1988).

It is crucial for an understanding of agricultural policymaking in later decades to note that the social democrats favored a policy that was based more on social than economic considerations. The three members of the royal commission who came from the farmers' organizations generally accepted the economic orientation of the report, as did their organizations in their official comments (*remissyttranden*). Opposition from farmers can be dismissed as a reason for modification of the commission proposal by the government. Also, the report was supported by the blue-collar union movement (Government bill (GB) 1947, no. 75; Niléhen 1984).

The position taken by the government can be understood in the following way. Many social democratic politicians and voters had an affinity to agriculture. A large number of the MPs still lived in the countryside though a decreasing number had farming as their occupation, as shown in Tables 2.1 to 2.4. The most important reason for the policy position for Agricultural Minister Sköld was the close association between the Social Democratic and Agrarian Parties. The idea of creating a coalition government continued to be a part of the tactical and strategical thinking of the social democrats. For example, an unsuccessful attempt to create a coalition government was made after the 1948 election (Jonasson 1972; Erlander 1973, 1974; Odhner 1988).

It should even be kept in mind that the social democrats romanticized the virtues of the small farmer, which prevented the party from supporting proposals for far-reaching structural rationalization of agriculture. Also, the party knew that support from the countryside would be necessary for electoral success. One of the earliest opinion polls conducted in Sweden showed that roughly 40 percent of the voters in agricultural municipalities in 1946 favored the social democrats but that the non-socialist parties were in the majority (Ingulfsson and Hagman 1950, 148f.). A political strategy to maximize rural votes could be undertaken since the interests of the consumers and the producers were not polarized at this time (Anderson 1987). The government bill was supported by a large majority in the Riksdag.

An administrative apparatus soon developed to specify and implement the general legislation. The National Board of Agriculture was put in charge of the policy goals of rationalization. County agricultural boards (*lantbruksnämnder*) were established as a subsidiary body in all 24 counties in 1948.

The national and county boards are governed by an executive board which includes representatives from the farmers' organizations, MPs, and experts. The county agricultural boards became responsible for internal rationalization (modernization) and external rationalization (mergers) of farms. The county agricultural societies were given the task of advising and informing farmers about crop and animal husbandry and operational mechanization. The societies are publicly supported interest organizations; with the 1947 decision their independence and authority diminished greatly (Tiderius 1977; SOU 1964:55). Other government agencies were involved in the rationalization effort as well.

The other primary goal of the 1947 agricultural policy was subsidies and support. They became the responsibility of the National Agricultural Marketing Board. The board was reestablished in 1950 and was given two very important tasks: setting the price of agricultural commodities and negotiating prices with representatives of the farmers. The Marketing Board supervises the quality control institutions for potatoes as well as eggs and dairy products. Both institutions were created in the 1950s. Among the members of the governing boards of the Marketing Board and the institutions are representatives of the farmers.

Other administrative changes occurred in the time period as the agricultural establishment was under formation. Farmers were much better organized, and their representatives were influential participants in the making and implementing of public policy. In this way, their interests became a part of government and politics. Their foremost spokesman in politics was the minister of agriculture.

The two policy goals from 1947 were implemented rather successfully. Incomes were equalized. Agricultural output increased while agricultural employment decreased. Between 1950 and 1960, there was an annual drop of 3.6 percent in agricultural employment, a higher decline than in most industrialized countries (Dovring 1965, 16). Modernization occurred, as farmers internally rationalized their farms with the help of the state. External rationalization went slower. Yet it can only be concluded that Sweden's transformation from a rural to an industrial nation occurred in this period.

Not everyone was happy with this situation. On the one hand, a realistic view of agriculture developed as farming created new problems for the national economy. Self-sufficiency was accomplished earlier than estimated. Surplus production became a serious and expensive problem. The price of food increased more than for other products. This led to the appointment of a royal commission with the task of studying the issue of price policy. A new calculation of commodity prices that gave more consideration to international prices and income comparisons was enacted by the Riksdag in the 1950s. As a consequence, the discrepancy between farmer and worker income increased. Forestry ownership by farmers was associated with farm rationalization. It became a controversial issue in this period as well.

On the other hand, the flight from the countryside in search of employment led to a new wave of romanticism. New media were used to deliver the message. For example, a very high proportion of the films produced between the mid-1940s and 1950s portrayed the countryside in a romantic way. City life was often scorned. The Agrarian Party used the rural myth as their foremost argument for state promotion of the combination of farming and forestry. The peak was 1948, when peasant romantic films composed a third of all films. The countryside continued to be a motif after the mid-1950s, but the romanticism had changed (Qvist 1986).

The Agrarian Party was affected by the trends as well. The declining number of farmers signified that the party would need to seek support from other groups. It changed its name to include the countryside in 1943, but it was not successful at the polls until the late 1950s. The Agrarian Party was influential in politics since it entered a coalition government with the Social Democratic Party in 1951. An important reason that explains the ocurrence is the Agrarian Party's desire to be able to exercise influence over how government would deal with the problem of surplus production.

The leader of the party, Gunnar Hedlund, considered the coalition government to be of great importance for agriculture. The agrarians headed four ministries, including the ministry of agriculture. Hedlund refused the position as its minister. It is reported that he thought it unwise for the party to be so closely identified with the interests of agriculture (Enochson and Petersson 1973, 79).

This concern did not prevent the party from exerting its influence over agricultural policy. A number of trade-offs were made with the social democrats. The interests of small farmers and those owning forests were safeguarded. Disagreement over agricultural and forestry policy created difficulties for the coalition government, but it did not contribute directly to the decision of the Agrarian Party to withdraw in 1957 (Ivarsson 1977, 175; Jonnergård 1984, 1985).

While the Agrarian Party continued to represent the interests of farmers, it sought to broaden its electoral base. An effort commenced in the late 1940s to gain support in densely populated areas. Peasant romanticism was deemphasized. After 1956 the party did not use campaign posters which focused only on farmers or people living in the countryside. In 1957, it changed its name to the Center Party (*Centerpartiet*) and began to emphasize the environmental issue. Officially the party no longer held a romantic view of agriculture, and fewer political decisions were motivated on the basis of their advantage for farmers (Larsson 1980; Ruin 1968; Elder and Gooderham 1978). The orientation proved to be attractive to voters, but agriculture did not cease to be a major concern for the party. Cooperation between the Center Party and the organizations of farmers is well-established.

AGRICULTURAL SUBGOVERNMENT

Acceptance of the exceptionalism of agriculture was achieved rather easily in Sweden. The farmers' organizations did not need to exert much effort to convince the other social groups and political parties that farming should be given special status. The main actors who formulated Swedish agricultural policy were sympathetic to the farm problem. Criticism of the policy under formation did not in any serious way affect the perception of the actors. Many of the critics were not part of the agricultural establishment, and they found it difficult to argue for a reevaluation of the farm problem for the main actors reasoned romantically.

It is noteworthy that peasant romanticism as an argument for exceptionalism was voiced more by the political parties than the organizations of farmers. The friendly and sympathetic political atmosphere toward agriculture would change in the years to come. The farmers' organizations began to formulate their demands in a more romantic way, as the predominance of farmers in the subgovernment was challenged by new groups demanding a role in policymaking.

It is intriguing to question whether the organization of farmers into interest organizations was necessary for the promotion of their interests in the period under review in this chapter. One answer is that their formation helped solve state problems, such as war preparedness and the administration of crisis policy. In later decades, the organizations would still prove to be useful for both the implementation of national policy as well as the representation and safeguarding of the interests of farmers. The predominance of farmers that is present in the first five decades of this century was modified gradually. Economic considerations would be given greater attention. Other political groups and problems would demand resources, and, most important, the interests of the consumers would be incorporated more fully into the agricultural subgovernment.

3

Swedish Farmers' Organizations

INITIAL ORGANIZATION OF INTERESTS

Unlike the situation for other interest organizations, the Swedish state has always played an active role in the interest organizations which represent agriculture. The Riksdag debated the establishment of a national organization for the furtherance of agriculture as early as the mid-1700s. The first organized groups gradually developed into "institutional groups," as they became characterized by a formal organization that performed political and social functions (Almond 1974).

Further efforts were made to organize the countryside after the Academy of Agriculture was created in 1811. County agricultural societies became more common and were present throughout the country by the mid-1800s. A few societies received public financial support in these early years. Some became involved in policymaking by submitting remiss comments to government proposals. The economic situation of the societies improved greatly after 1855, for they began receiving a portion of the tax on aquavit.

The change that agriculture went through at the end of the last century encouraged the development of other organizations as well. The purchasing of crop seed and the introduction of commercial fertilizers are directly related to the decision by farmers to form joint purchase cooperative societies.

Cooperative dairies appeared in the 1880s. They comprised 5 percent of all dairies in 1890, 25 percent in 1900, 39 percent in 1910, and 45 percent in 1930. Now they have almost complete control of the market, a development that is seen as unfortunate by the state agencies established to safeguard antitrust policy. Egg and meat marketing cooperatives formed as well. While egg marketing cooperatives had about 14 percent of the market by 1930, those for meat marketing held 40 percent of pork and 4 percent of beef sales.

The explanation for the organizing efforts in these years is threefold. Farmers realized that certain private wholesalers took advantage of their situation. They wanted to be guaranteed that the quality of their own purchases was high. They desired to buy goods at the lowest price possible. Other organizations for agriculture and the countryside were also created at this time.

Not all cooperative societies were successful. An important event in 1895 that changed this was the enactment of a law creating a new legal category of business association, the economic association (*ekonomisk förening*). The law promoted the growth of cooperative organizations by stabilizing their financial situation. Inspiration came in particular from the successful experience of the German cooperatives. The county agricultural societies were actively involved in all organizing efforts of this period. Also, the societies grew in strength and influence and were considered at that time to be the sole representative of the agricultural interests in Sweden (Jo 1921, 27).

The first national farmers' cooperative association appeared in 1905. It was the Swedish Farmers' Supply and Crop Marketing Association (*Svenska Lantmännens Riksförbund,* SLR), an organization supported by both small and large farmers. By 1919, it organized farmers whose holdings comprised 15 percent of all arable Swedish land; by 1929 it was 26 percent, though the regional distribution of membership was uneven (SLR 1955, 66ff.).

The food crisis of World War I changed the situation for the farmers' organizations. As the only national organization of its kind, the Swedish state depended on the SLR to help solve the problem of food distribution. The SLR used its close contacts with state agencies to help the farmers, a circumstance that lead to success in member recruitment.

Though the wartime crisis encouraged farmers seriously to consider collective action, only some of the cooperatives which existed prior to the war grew in strength. Others stagnated, due to the severe economic situation. A national farmers' association that brought together individual farmers and the different organized groups, including the county agricultural societies, was established in 1917. Its name is the Swedish Agricultural Society (*Sveriges Allmänna Lantbrukssällskap*, SAL).

The SAL became a trade association for agriculture, which involved interaction with the state on policy issues. Yet the organizational effort did not lead to the creation of an effective national association that articulated the interests of farmers. Within four years, its membership shrunk and economic situation declined. The development was seen as lamentable in both government and agricultural circles as the agricultural crisis of the 1920s worsened.

The SAL was reorganized into an association for the farmers' cooperatives. Two main tasks were the extension of the cooperative movement into new fields and recruitment of existing farmers' cooperatives as members. With the help of the SAL, five new national cooperative associations were formed in the early 1930s (*see* Table 3.2 for more information).

A decade later, the SAL underwent another reorganization. It became the peak or central organization for the cooperatives, which signified that direct or union membership was no longer possible. Other organizations, the county agricultural societies for example, could no longer become members. Its new name, the National Federation of Agricultural Cooperatives (*Sveriges Lantbruksförbund,* SL), emphasized more its character as a general interest organization than coordinator of the cooperative movement.

A second important organization in the history of the Swedish farmers' movement was founded in 1929. The point of departure for the National Swedish Farmers' Union (*Riksförbundet Landbygdens Folk*, RLF) was the belief that solidarity and collective action was a necessary strategy for the betterment of the economic situation of farmers. Trade unionism inspired its founders. Its aim was to bargain collectively on behalf of the farmers.

Discussion in the early stages centered on whether disciplinary action could be used to ensure solidarity and whether the RLF could organize members who were neither farm owners nor tenant farmers. The two points are currently on the agenda of the present farmers' organization. The RLF decided that forest owners and people living in the countryside should be offered membership. The name of the organization reflects this recruitment policy: a literal translation into English would be the "National Association for the People of the Countryside."

Two years after its creation, the RLF had 20,000 members. Ten years later, the number jumped to 68,000. One reason for the increase is the incorporation of a smaller conservative and militant farmers' union organization into the RLF. The RLF became an influential popular movement that was supported by an enthusiastic corps of small farmers.

Its relationship with the SAL, later the SL, soon stabilized. Action taken by the RLF is an important reason for the creation of many farmers' cooperatives and their ability to recruit members. By the time the agricultural crisis had passed, farmers were much better organized than at any earlier time. One consequence of organizational growth was a more coherent articulation of interest. Different categories of farmers found that they had interests in common. Also, the farmers were well-represented by a rather united farmers' movement that included a union, a peak association for the cooperatives, and a political party (Bjärsdal 1983, 1986; Thullberg 1977; Johansson and Thullberg 1979; Micheletti 1987a).

ROLE OF STATE IN ORGANIZATIONAL DEVELOPMENT

The political attention given to agricultural policy in the 1920s and 1930s can in part be explained by the pressure that the burgeoning farmers' movement put on the state. Both the SAL and RFL sought financial support for their organizational activities. The SAL received support for its two main tasks to strengthen the cooperative movement that are discussed in the previous section. The main interest of the state was the development of the farmers' cooperatives into effective business enterprises. Effectivity meant centralization and rationalization. The state considered the SAL a better vehicle for this. In fact, the 1928 reorganization of the SAL was designed to attract the attention and obtain the acceptance of the government.

Public finances to support the cooperative movement were allocated for the first time by the Riksdag in 1930. Two cooperative associations, the Swedish Dairies' Association (*Svenska Mejeriernas Riksförbund,* SMR) and Federation of Swedish Cooperative Banks (*Sveriges Föreningsbankers Förbund,* SFF), were established with the help of the support. That the government promoted cooperative businesses is not an uncommon occurrence when comparisons are made with the actions of other nations in this period. Nor is it unusual that dairy cooperatives were given priority (Manchester 1983).

The experience is interesting as it exemplifies a general type of relationship between interest organizations and the state in Sweden that developed in the years of crisis and that is prevalent today. Understanding of the historical development of agricultural policy helps explain the present policy dilemma. The relationship has three general characteristics. First, the interest organizations are delegated a number of specialized and crucial administrative tasks by the state. The tasks require detailed practical knowledge of the policy issue and a good rapport with the groups that are their targets. Second, the organizations become more closely involved with the state as it becomes more active in its policy of intervention. A policy sector as well as a subgovernment emerge. Third, the state and the interest organizations both become a spokesman for the interests of the targeted group, in this case farmers (Lewin 1967, 177; Hellström 1958, 55, 88; Kristersson 1954, 10; Herlitz 1936, 228; Reuterskiöld 1936).

Financial support was accepted by the national associations of agriculture. Member discipline was encouraged by both the organization and the state. The state was allowed to participate in the formulation of organizational tasks, goals, and structures. As a consequence, it became both more receptive to pressure group demands and interwoven in organizational problems.

The case of the dairy cooperatives illustrates this development well. The SMR was created in 1932 and was that year delegated by the Riksdag the responsibility of regulating the milk market. The proposal for milk regulation was formulated by the SL's dairy unit (later to become the SMR) and submitted to the government. A number of cooperative dairies were in existence but, as shown earlier, they did not have a large share of the market. The proposal involved a fee on milk to be collected by a national association from all farmers whose milk was delivered to dairies, regardless of whether the farmers were members or not. The idea was that the stabilization of milk prices required central administration and support from milk farmers. Most farmers had dairy cows at the time. It was argued that voluntary compliance would not solve the problem.

The Riksdag agreed with the basic proposal from the SL. The SMR was created, delegated broad responsibilities by the state, and received public finances for its tasks as a public agency (Friberg 1973; Åkerman 1937; Ekegren and Hagander 1938, Ringborg 1940; Heckscher 1951, 119ff.). One year after enactment of the government bill, the SMR could boast that it organized 882 cooperative dairies that received 77 percent of all milk delivered to dairies. By 1939, the figures were 887 dairies with 93 percent of the market share (LRF 1980, 44f.).

Other farmer cooperatives, the Swedish Meat Marketing Association (*Sveriges Slakteriförbund,* SS), the Swedish Egg Marketing Association (*Svenska Ägghandelsförbundet,* SÄ), and the Swedish Starch Producers' Association (*Sveriges Stärkelseproducenter,* Stärkelsen), were also delegated public authority, and public finances were used for the task of membership recruitment (Ringborg 1940; SLR 1955; Seyler 1983). The market monopolies that developed were not viewed as a negative development until the 1960s.

Sweden coped effectively with the problems of food supply and distribution in World War II. An important reason that explains the improved situation is the role played by the farmers' movement, which had grown in strength in the inter-war years. The RLF had 76,300 members by 1940.

Farmers also joined the farmers' cooperative movement; approximately 720,800 memberships were registered that same year.

The RLF became a general interest organization for farmers that was characterized by its high levels of member activity, solidarity, and service to members. Contacts with members were good. It evolved into an important spokesman for the interests of farmers, particularly on political issues and in public opinion formation endeavors, while the SL had developed into an effective business operation with good contact and rapport with the state.

Both organizations took the initiative to request a negotiation with the government on the setting of agricultural prices after the bad harvest of 1940. An important reason for it being granted was a referendum among the RLF members which showed that they were willing to strike for better prices. It is the first time a negotiation took place. Another referendum among the members on price-setting was held in 1945 (Thullberg 1980; Hellström 1964).

The RLF was somewhat more hesitant than the SL to negotiate with the government. The hesitation is explained by its character as a union organization for farmers. A negotiated decision in 1942 that dealt with the effect of milk prices on inflation policy met with the displeasure of a number of regional units within the RLF. It is noteworthy that they refused to grant the RLF leadership the right to approve the agreement made with the government. The war situation otherwise encouraged consensus, understanding, and discipline within the memberships of different interest organizations.

Two positions were taken by the government on milk prices. The social democratic minister of finance wanted to heed the opinion of the Confederation of Swedish Trade Unions (*Landsorganisationen i Sverige,* LO). The agrarian minister of agriculture opposed a lower milk price and threatened repeatedly to resign.

Though highly criticized by the Liberal and Conservative Parties, leaders of the Agrarian Party, public agencies, and interest organizations within the agricultural establishment, farmers within a RLF local unit threatened to strike for higher milk prices. The strike action, which occurred in the province of Dalarna, put the Agrarian Party in a difficult position, as it was seriously divided on the issue. Neither did the SL give its support.

The RLF leadership was also in a very uncomfortable position. If it were to support its members, it would lose the trust of most political actors, be shunned by the agricultural establishment, and be barred from further contacts with the government. If it were to accept the negotiated proposal, it would alienate many of its members. The leadership had in reality two bargaining partners—the agricultural establishment and its own members.

The action taken by the RLF leadership offers an excellent example of the practical working of Swedish political culture at its best. Conflicts that cause severe economic and political damage are avoided. Integrity and honor remain intact. Problems are solved. Winners and losers of conflicts are not proclaimed.

The leadership decided not to give explicit and public support to the threatened strike actions. Further, it requested that the government negotiate anew on milk prices. This occurred, and milk prices were subsequently raised.

The government modified its position. It became obvious that the protesters had the wide support of farmers from other regions. The social democrats feared that a hard line might split the Agrarian Party and jeopardize the coalition government. The government was aware that farmer discontent could conceivably be channeled through a fascist movement and also used as an argument by the communists. The leaders of the RLF showed the government that they were able to assume public responsibility. Four years later, the RLF adopted a declaration of principles in which, among other points, it was stated that more responsibility for member discipline would be accepted (Lewin 1967, 177; Thullberg 1980, 1988; Johansson and Thullberg 1979).

Coordination and cooperation between the state and the farm organizations was as important in later years. Involvement of the organizations in the farm rationalization programs was seen as essential. The interest organizations could perform certain administrative tasks. Together with the quasi-public county agricultural societies, the SL and RLF were both engaged in the extensive reform that agriculture underwent in the postwar period.

POLITICAL PARTIES AND FARMERS' MOVEMENT

Swedish political parties have supported the organization of farmers into associational groups. Associational groups are organizations which represent the expressed interests of a particular group, employ full-time professional staffs, and have orderly procedures for the formulation of interests and demands (Almond 1974). A positive view of the mobilization of farmers came from political parties whose political ideology did not postulate farmers as a natural ally. They includes the social democrats and other parties of the left in the early decades of this century.

Their encouragement was in part a reaction against the more ideologically conservative farmers' organizations. A leading social democrat proposed in

1920 that small farmers form a union organization to articulate their particular interests and raise their class consciousness. Explicitly he criticized the other farmers' organizations existing at the time. The SAL and the county agricultural societies were seen as representing only the larger farmers and as based on elite democratic principles (Thullberg 1974). The Social Democratic Party supported the farmers' movement in the years to come for different reasons.

One reason is the generally positive attitude of the party to interest organizations that help less articulate citizens further their political and economic interests. A second is the historical compromise and the years leading up to it, which implied that the party dealt favorably with the farmers' organizations as they were intimately associated with the Agrarian Party.

Social democratic attempts to negotiate an agreement with the Agrarian Party date from the early 1920s. The Agrarian Party used the opportunites to help the farmers' movement advance politically. In a negotiating session in 1948, it successfully proposed that the SL and RLF be given greater influence over agricultural policy, particularly farm rationalizations, and in the executive boards of the county agricultural boards (Jonasson 1972; SLR 1955; von Heland 1969).

A third explanation for the support given to these organizations is the historical development of Swedish corporatism. Finally, it may be argued that the social democrats wanted to support cooperatives as a business form. The presence of cooperative enterprises in agriculture was an important argument against its nationalization (Odhner 1988).

Whether the farmers' organizations would have played a different role in politics without the presence of the Agrarian Party is a question which cannot be answered. It seems that the party was very important in creating legitimacy for the RLF and SL. But the organizations were capable of exercising influence over the state themselves. The question has current relevance. The Center Party desires to evolve from its agrarian roots, and the LRF has recently decided to recognize this fact and seek support from all political parties. There is no doubt that the intimate relationship between the SL and RLF, later the LRF, and the Agrarian Party, later Center Party, has benefited Swedish farmers, especially those with larger holdings.

The relationship between the Agrarian Party and the farmers' organizations has historically been characterized by a network of close personal contacts. In their formative years, the identification of certain leading politicians whose policy speciality was agriculture with the interest organizations proved to benefit the farmers' movement, as the state entrusted it with public responsibility without demanding more public control.

Close association with the party in the form of party leaders as members of the executive boards of the national associations and peak organizations was considered important for the cooperative movement which was highly dependent on state agricultural policy. Until the late 1960s, political experience was almost an essential qualification for the SL and RLF board members. Yet it may be argued that both the SL and RLF realized at an earlier date that they could exercise more influence in politics through their expertise than through their close contacts with the Agrarian/Center Party.

The party and the organizations have historically relied on each other as a recruitment base. Organizational leaders have held influential posts in the government ministries, and party leaders have become presidents and directors of farmers' organizations. Leading spokesmen from the organizations have been elected to the Riksdag, and many have worked as under secretaries of state. They have in some cases represented parties other than the Agrarian/Center Party, including a conservative minister of agriculture. Certain farm leaders have ended their careers by becoming county-governors and holding posts within the agricultural establishment as director-generals.

A clear association with the Agrarian Party is historically evident among the active memberships. A study conducted in the early 1950s showed that 55 percent of all members of the executive boards of the farmers' cooperatives were active party members; another 10 percent were well-known as party supporters. The situation is similar for the RLF: over 80 percent of the board members participated actively in the Agrarian Party, and an additional 7 percent were well-known as sympathizers. The Conservative Party received more sympathy within the SL cooperatives (20 percent) than within the RLF (8 percent). Two other parties whose support among voters in the countryside was higher than reflected here were the Social Democratic and the Liberal Parties. Sympathy for the Agrarian Party among the elected leaderships was higher than the agrarian votes cast for it (Elvander 1972; Hellström 1964, 90—96).

Political controversy played a role in organizational elections, but its significance for internal politics should not be exaggerated. The RLF had fewer problems with political controversy as the Agrarian Party dominated its leadership. The SL and RLF were both eager to emphasize that they were organizations of political neutral character. Certain leading agrarian politicians held the same opinion. A more formal structure for contact between the party and the organizations similar to the labor movement was proposed but never realized. It is interesting that the logotype of the RLF and the Agrarian Party (and later the LRF and the Center Party) are quite similar. Both use a leaf of clover.

The SL and RLF, and later the LRF, have numerically dominated within the Agrarian Party. Neither did the transformation of the Agrarian Party to the Center Party seem to affect the strong bond, as confirmed by available statistics from different time periods. MPs for the Center Party in the 1963 Riksdag were overwhelmingly members of the SL (84 percent) and the RLF (93 percent). Many of the MPs considered themselves to be spokesmen for the farmers' organizations. Organizational membership was not present in any other party to the same degree. Membership in blue- and white-collar unions among social democratic MPs was roughly 70 percent. A considerable number of the Center Party's MPs were active members in the farmers' movement.

Compared to most other major interest organizations, the farmers' organizations had a greater number of elected officials as MPs. Members of Parliament from the Center Party were also found to be more open to the lobbying efforts of interest organizations, and the farmers' organizations were allowed to discuss political issues directly with the party's policymakers in the Riksdag (Elvander 1972, 107—12).

Personal contact and interchange affected the policies promoted by the party and organizations. Though the subject has not been investigated systematically, research shows that an exchange of influence occurred. The RLF was, for example, involved in the negotiations on the formation of a coalition government in 1932, 1948, and 1951. In exchange, the RLF announced to its members that it approved the agreements that were reached. It is noteworthy that the RLF's presence in the 1948 negotiation is the official reason for the participation of the LO, as emphasis was placed on the importance of balancing the involvement of the participating interest organizations (Thullberg 1977, 282—93; Ruin 1968, 221, 373).

The RLF exercised influence over leaders of the Agrarian Party. It has been reported that pressure was put on the party leader in the coalition government of the 1930s. One source states that the party leader was considered to have represented the interests of farmers inadequately, and many leading RLF spokesmen demanded his resignation. The conflict was solved when he promised to follow a decision made at the party congress that outlined the conditions for the involvement of the party in the coalition government (von Heland 1969, 140ff.).

Mutual reliance or pressure characterized the cooperation in later years. The campaign that was sponsored by the RLF on the greening of Sweden in the 1968 parliamentary election helped the Center Party, which desired to focus more attention on its special issues: the environment, decentralization, and the advantages of small-scale organization. The RLF and the Center Party agreed on how the campaign was to be formulated, an agreement which ignored the contradictions between the party's message and the realities of agricultural production, that is, its large-scale production and reliance on chemical pesticides.

It seems also reasonable to assume that cooperation between the RLF and the Center Party in this election had a more long-term effect in that it aided the party as it attempted to transform itself into an environmental party attractive to urban voters. One political commentator compared the cooperation in the electoral campaign with the traditional cooperation between the Social Democratic Party and LO (Olivecrona 1968, 134f.).

More recent investigations present results showing that the LRF is not as important politically as its predecessors. One study found that slightly over one third of the MPs from the Center Party in the 1985 Riksdag were LRF members (Gustafsson 1986). Another scholar researched the organizational affiliations of the delegates at the 1983 Center Party congress and found that membership in the LRF was still the most frequent affiliation but that the interest of farmers does not dominate the party (Uddman 1986).

The implications of declining membership rates among politicians from the Center Party for the organized farmers has not been investigated directly. A more comprehensive study of MPs in the 1985 Riksdag sheds some light on the issue. It found that organizational affiliation correlates positively with concern MPs show for the interests represented by the organization. The study confirms the decreasing influence of farmers over the Center Party: 15 percent of its MPs in 1985 considered it important to represent the interests of farmers (as compared to 21 percent in 1969).

It is intriguing that an increasing number of MPs for the Conservative Party view farmers as an important group: 17 percent in 1985 compared to 10 percent in 1969. The majority of the members of the LRF continue to vote for the Center Party, though the proportion has decreased from 77 percent in 1976 to 69 percent in 1985. More vote conservative today (Holmberg and Gilljam 1987, 188; SCB 1970, 100).

Yet it can be concluded from the study that farmers are still more influential among MPs than they should be, given their proportion of the Swedish population (about 3 percent). In 1969, 6 percent of all MPs considered their interests as important. By 1985, 8 percent thought so. The increase is explained by the growing attention given to farmers by the Conservative Party (Holmberg and Esaiasson 1988, Tables 2.1—2.3).

Concern among conservative MPs for the interests of farmers has not meant that contact between the LRF and the Conservative party has increased to the detriment of the Center Party. More Center MPs are in contact with the LRF than with any large interest organization. The rates of contact between the LRF and the political parties of the Riksdag are: the

Left Party Communists (*Vänsterpartiet Kommunisterna*, VPK), 11 percent; Social Democratic Party, 4 percent; Center Party, 32 percent; Liberal Party, 8 percent; and Conservative Party, 8 percent (Holmberg and Esaiasson 1988, Table 3.1). Contact between the LRF and Center members of the Standing Committee on Agriculture in the Riksdag is also high, though the LRF tends to have a high degree of contact with all members of the committee (Holmberg and Esaiasson 1988, Table 3.2, 71f., Verba et al. 1987, 64).

RELATIONS WITH OTHER INTEREST ORGANIZATIONS

The two farmers' organizations that are the main subject of this chapter have had numerous contacts with many different interest organizations. The contacts that possibly have been most constant are those with the Swedish countryside, a geographic area and interest complex that is saturated with various interest organizations. In most cases, the RLF and SL (later the LRF) are members of the organizations for the countryside and view their activities as an important complement to the tasks performed by the farmers' movement. One way to classify the rural associations is after their main organizational purpose.

Certain ones concentrate on providing agricultural advice and service. Foremost among them are the county agricultural societies. They are now organized in a national association with over 100,000 members and 1,000 employees. The societies developed in an interesting way, as they have been more affected by state involvement in agriculture than any other farm organization. Intimate contact between them and the RLF occurred at the end of the 1960s, a time when both organizations experienced economic hardship. Contact continued later with the LRF. The role played by the societies may be enhanced in the near future, as a consequence of the reform of the Swedish public sector.

A second category of organizations are educational. Some offer specific courses and programs on farming. Others concentrate more on general areas of study, for instance the role of agriculture and the countryside. This latter type, the adult education association (*studieförbund*) is a classical Swedish organization and important institution in Swedish political culture.

The association that the LRF is involved with is the *Studieförbundet Vuxenskolan* (SV). (There exists no official English translation.) It was created in 1967 after an amalgamation of the adult education associations for the Center and Liberal Parties, one example of the close cooperation between the parties in this period. The RLF and SL were members of the Center Party's association. Until 1967, the Cooperative Union and Wholesale Society (KF) was a member as well.

Initiative for the creation of an adult education association was taken by the youth organization connected to the Agrarian Party. As has often been the case when Swedish farmers have decided to organize their interests, they wished to pattern their proposed association after the one created by the labor movement. The history of the agrarian adult education association is interesting as it reflects both the development of agrarian thought from individualism to collectivism and the politics of the times.

While education on agricultural economics was stressed in the 1920s, emphasis was put on peasant romanticism and nationalism a decade later. By the 1950s, regional politics, the culture of the countryside, and the organized interests of agriculture became important. They were also high on the political agenda. The SV's education program from 1979 gave less attention to the farming population and toned down the participation of farmers' organizations, which coincides with the efforts of the Center Party to mobilize strength in urban areas (Johansson 1985, 70—78).

Another category of rural association in contact with the farmers' organizations engages in opinion formation activities and contacts with government. The local culture society, mentioned in Chapter 2, is illustrative here. Often opinion formation is included in the activities of associations that articulate the many different special interests in the countryside. There is a wealth of them, including special organizations for women, the youth, and estate owners.

A few organizations that represent agricultural interests compete rather than cooperate with the farmers' movement. They have not been important as a force for individual farmers or in discussions on agricultural policy with the state.

How the relationship between the farmers' and consumers' cooperatives has developed is interesting. While the movements represent diametrically opposed interests (consumers versus producers), they not only constitute a form of enterprise that differs from both nationalized and private business but are primarily active in the same market, the food sector.

Initially farmers helped consumer cooperatives establish themselves in the countryside. The consumer cooperative movement was soon more advanced than the farmers' movement. Some authors conclude that the consumers' cooperative was the largest organization of farmers at the outbreak of the agricultural crisis in the 1920s, as its membership included around 100,000 farmers and farm workers (Kylebäck 1984; Stolpe 1976).

A regional farmers' supply and marketing cooperation belonged at first to the KF (established in 1899), but the creation of the SLR in 1905 signified that the development of one cooperative movement for both consumers and producers would not be realized. Cooperation rather than competition characterized the relationship between the KF and SLR in the early years. An unsuccessful attempt by the SAL was made in the 1920s to establish a general coordinating body for the entire Swedish cooperative movement.

As the two cooperative movements grew in strength, they began to compete, in particular over the dairy and meat market. The KF considered the support that was given to the farmers' cooperatives by the state in the 1930s to have put the farmers in an advantageous position. In 1936, an agreement was made between the SAL and KF regarding their respective spheres of business activity; another one was signed in 1945. The conflict between the two cooperative movements meant that the SAL decided in 1935 to use the term "economic associations' movement" (*ekonomiska föreningsrörelsen*) rather than the term cooperative (*kooperation*) to describe its organization. Thirty-eight years later, the term "farmers' cooperatives" (*lantbrukskooperationen*) was again used officially.

Competition and artificial coordination rather than spontaneous cooperation and natural affinity tends to characterize the relationship between the consumers' and farmers' cooperative movements. The KF has criticized one of the main principles that is applied within the farmers' cooperation, the rule of obligatory delivery of products to the cooperatives (*leveranstvånget*). It has also become an official counterweight to the farmers' movement in that it represents the interests of the consumers in the agricultural establishment. Its main point of view is that the interests of consumers should be the main focus of agricultural policy.

The LRF has provoked the irritation of the KF on a number of occasions. At the same time, the movements coordinated their purchase of petroleum in a joint organization until 1987 when the KF bought the LRF's shares, and an employers' association exists for dealings with unions whose members are employed by the cooperatives. There are other coordinated endeavors. One which also includes the state involves advice and financial support to people who wish to start cooperative enterprises.

A similar kind of relationship can be said to describe contacts between the LRF and the two foremost peak or central unions, the LO and the Central Confederation of Professional Employees (*Tjänstemännens Centralorganisation,* TCO). The unionization of farm laborers in the 1930s was not seen as a threat by the farmers' movement, as it did not attempt to recruit them itself. Also, the farmers' movement viewed the labor movement as a model of organization. Presently both the TCO and LO include among their members respective unions for agricultural white- and blue-collar employees, and the LO contains a special union for those employed in the food processing sector.

As representatives of the consumers' interests, the LO and TCO have, along with the KF, become part of the countervailing power that the organized farmers must face in agricultural policymaking and the price-setting negotiations. Leaders of both the unions are vocal in their criticism of the current agricultural policy. Food prices and the costs of regulation are seen as directly related to the economic situations of their own memberships.

A category of organization is becoming increasingly important for the farmers' movement. It is its international contacts: Nordic, European, and international associations of farmers and cooperatives. An example is the Central Council of the Nordic Farmers' Associations (*Nordens Bondeorganisationers Centralråd,* NBC), established in 1934. It is a forum for exchange of information on organizational activities and developments in agricultural policy. The LRF is the only Swedish farmers' organization with membership in the NBC.

The present farmers' cooperatives are members of the employer associations that are included in the Swedish Employers' Association (*Svenska Arbetsgivareföreningen*, SAF), the peak association for private employers. Involvement in guild or trade associations is deemed relevant as well.

REORGANIZATION AS RESPONSE TO CHALLENGES

Many changes occurred since the creation of the RFL and SL. Two world wars were fought and two food crisis situations weathered. One consequence was active state intervention in agriculture. Swedish agricultural policy promoted more modern production. Due to farm rationalization, farms grew in size and farmers decreased in number. Industry needed workers, and the ensuing decline in the farming population had an immediate impact on the RLF's membership. Also, farmers became specialized. Product specialization meant that the all-round farmer disappeared, and this affected the farmers' movement by dividing farmers into basic but different categories of interests, those specializing in cattle, hogs, hens, or crops. The specialization has significant impact for member solidarity and organizational structure.

Mention should be made of other changes. The farmers' cooperative movement became big business in the 1960s. It was vertically and horizontally integrated, a process which continues to this day. Mergers were seen as economically and technically necessary for the dairy, meat marketing, as well as grain marketing and purchasing cooperatives. To the dismay of a vocal minority, Swedish society became more centralized and large scale in the 1960s. The wave of mergers from 1960 to 1973 within the farmers' cooperatives met with criticism.

Not only did the farmers' share of the food market begin to be characterized as monopolistic in the public debate. Many members of the RLF themselves held a negative view of them and the centralization within the cooperatives caused by them. As sympathizers of the Center Party, they supported decentralization and small-scale production. The party encouraged them to voice their criticism within the RLF (Utterström 1980, 172).

Democracy within the cooperatives became problematic because of the mergers. Impersonalization, bureaucratization, and complications were the result. The mergers of the 1960s created often heterogeneous membership bases for the economic associations, causing conflicts between different groups of members. One study concludes that while democracy within the cooperatives was not affected in any considerable degree as member attendance did not decrease after mergers, member influence changed. The influence that members could exercise over important decisions within the cooperatives decreased. Yet larger size signified that more influence could be exercised over the market and political environment (Utterström 1980). It is an example of a classical problem for democratic institutions. As organized groups grow in influence, individual members lose influence over them.

The creation of larger cooperatives led to organizational professionalization and independence, which in turn caused a decline in the influence that the SL could exert over its member associations. Mergers made other problems more apparent. One example is the conflicts between active minority groups and the passive majority of members (Utterström 1980). Societal development and new patterns of food consumption affected this branch of the farmers' movement as well.

The SL was in the 1960s economically sound but had problems with its democratic structure. The situation was the opposite for the RLF. It was so severe that the RLF feared that it would prove devastating in the near future. Financial problem stemmed from Swedish agricultural policy. Specialization of production made the RLF's system for calculating member fees outmoded, as it was based on the number of acres owned or rented rather than on the productivity of the acres; farm rationalization caused membership rates to decline from 192,145 in 1960 to 133,000 in 1970. The crisis of the farmers' movement was acerbated by the general political debate on agricultural policy (Persson 1983; RLF and SL 1969).

The question was whether the RLF and SL should not attempt to solve their respective problems jointly. The issue had been broached in the past, and joint endeavors were not uncommon. An important example is their delegation for negotiations which was established in 1948 for the formulation of a united farmers' interest in negotiation on commodity prices with the state. The RLF supported the cooperative movement and encouraged its members to join it. A joint council was established in 1935 to coordinate the activities of the two organizations, though it was never very influential.

In 1938, the SAL conducted an investigation that proposed a new peak or umbrella interest organization for farmers with the economic associations, county agricultural societies, the SAL, and RLF as members. The proposal was not accepted. Two years later both organizations agreed to use the same statistics as a basis for the first price-setting negotiations with the government. A contract was signed in 1946 which obligated both organizations to consult each other so that different opinions were not put forth in negotiations with the state. They were also granted a seat on each other's executive board; cooperation between their respective newspapers was established, and a joint research unit was founded in 1950.

It is often claimed that true agreement on intentions, opinions, and strategy does not need to be clothed in formality. That the attempts were not particularly successful became very apparent in the 1960s, a decade that changed Swedish agriculture and challenged the farmers' organizations. A royal commission was appointed to reevaluate Swedish agricultural policy. The organized interests of the consumers became a countervailing power in agricultural policymaking. The new law on property ownership from 1965 made it easier for legal persons to purchase farmland and promoted farm rationalization. The SL and RLF considered the proposed law a threat to farming and the countryside. They wrote a joint remiss reply and coordinated their attempts to influence the political parties.

Coordination did not signify that the RLF and the SL were necessarily in agreement in these years. Organizational leaders generally agree that the challenge to agriculture in this period strained the relationship. The basic disagreement stemmed from the fact that the organizations represented different interests of agriculture: the farmer as owner, laborer, and resident of the countryside versus the cooperatives as a business enterprise.

It became difficult for the organizations at the national level to coordinate their standpoints in discussions with the state. The outward appearance of interorganizational agreement often masked disagreement and struggles over influence and status. Organizational leaders and members, politicians, and the mass media began to wonder which organization represented the farmers (Persson 1983). Reorganization became a very attractive alternative when consideration was given to all the problems that faced the farmers' movement.

The first formal discussion on the relationship between the two organizations was in 1963 but did not lead to any change. A new investigation was appointed in 1966 and concluded its task in 1968. Three different alternatives ranging from more coordination to amalgamation were presented. A number of the RLF spokesmen feared that the cooperative enterprise would dominate the farmers' union if the organizations were to amalgamate. Whether the union's right of strike could be maintained after amalgamation was an important issue. There were technical organizational problems that needed to be solved as well.

In a rather recent interview, one leading spokesman maintained that the issue of increased cooperation between the two organizations was crucial because the RLF and SL were finding it more difficult to come to terms with each other. "They were," as he stated, "on a collision course" (as quoted in Persson 1983, 34, my translation). The problems facing the agricultural sector were of such magnitude that united effort was deemed as essential by many leaders. Yet most of the RLF leaders opposed an amalgamation, and information to the members on the proposal was delayed.

A compromise was reached and from 1970 the farmers had a new organization, the Federation of Swedish Farmers (*Lantbrukarnas Riksförbund,* LRF). The official reasons for the reorganization emphasized the issues of efficiency and effectiveness. Attention was focused on the name of the organization. That the name was similar to the RLF and that the RLF logotype would still be used was considered to be important for symbolic reasons. Many leaders and members of the RLF distrusted the cooperative movement and feared that it would engulf their organization and stifle its democracy and spirit of solidarity. Preservation of the logotype was a symbolic attempt to hinder this development. In the early years of LRF, criticism against the modernization of the logotype was voiced at the General Annual Assembly (Persson 1983; LRF minutes 1971).

FEDERATION OF SWEDISH FARMERS

Organization

The Federation of Swedish Farmers is not, in a strict sense, a true amalgamation, as it is not characterized by a complete merger of the RLF and SL. Rather the two organizations have generally retained their former structures. Diagram 3.1 depicts the LRF organization. It is noteworthy that the union and cooperative branches are easily distinguised as their local units have been retained.

Shortly after the creation of LRF, many members claimed that they did not feel the effects of the reorganization. Though preservation of the traditional structures for collective action was initially seen as necessary, the incompleteness of the amalgamation is a recurrent topic at the general annual

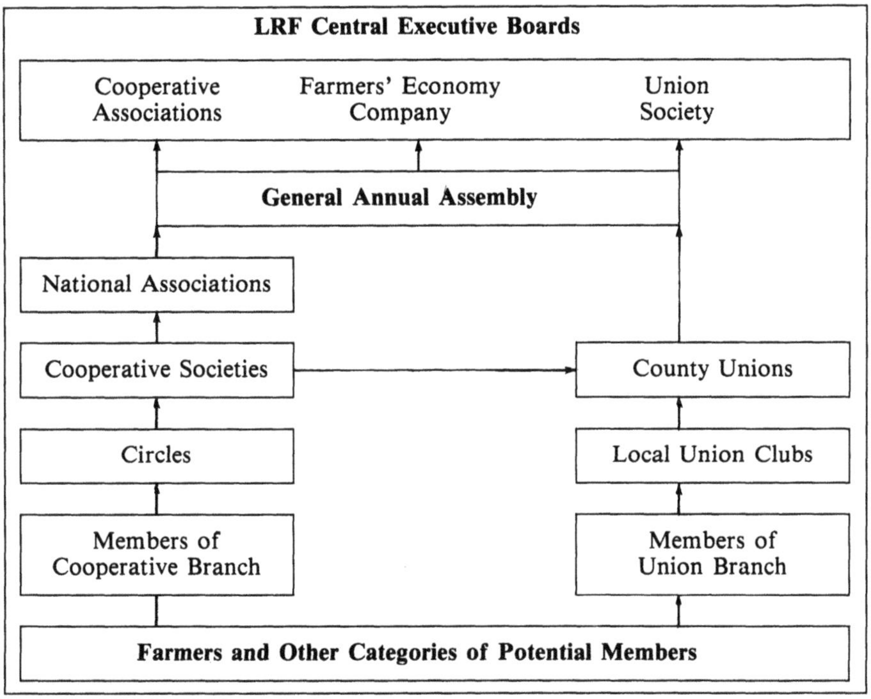

Figure 3.1. Organization Structure of Federation of Swedish Farmers

assembly and in the discussions within the organization. Greater coordination and cooperation between the two branches is called for repeatedly. Change is now underway.

Members of the local union clubs (*lokalavdelningar*) are farmers, their families, people whose employer is the LRF, residents of rural areas, and sympathizers of the farmers' movement. The number of nonfarmer members has increased over the years. Recruitment of that category is emphasized currently, as the LRF attempts to broaden its membership base to include the entire countryside. Neither did the general tasks of the union clubs change after the reorganization, though they have gradually become more numerous to include at present various union matters and agricultural, forestry, regional, tax, property, and industrial policy. Recruitment of members, opinion formation, and debate as well as the representation of members on governmental matters at the municipal level are other activities of importance. Geographically, the union clubs cover one or two parishes. Though they include a greater number of members than in the past, they are still small units. Many have less than one hundred members. Table 3.1 offers statistics on the union clubs and membership in the union branch of LRF.

Table 3.1
Membership in the LRF Union Society and Number of Local Clubs, 1972—1988

Type of Membership[1]	1972	1975	1980	1985	1988
Farmer	122,860	126,119	131,922	132,056	124,501
Nonfarmer	9,800	14,540	19,421	23,327	23,163
Total all members	132,660	140,659	151,343	155,383	147,664
% farmer	93	90	87	85	84
% nonfarmer	7	10	13	15	16
% total	100	100	100	100	100
Local Clubs (N)	1,900	1,730	1,684	1,650	1,650

Source: LRF, *Lantbruksåret* 1987, LRF, unpublished statistics.
1. *Farmer members* are those who own or rent farms and their family members. *Nonfarmer members* are those who actively sympathize with the LRF by paying membership fees and their family members. They do not have farming as their occupation. The category includes the officials employed by the LRF.

In the past, the local union clubs were essential for the type of democracy that was encouraged within the RLF. A study conducted on interest organizations and popular movements in the 1960s found that the tendencies toward member alienation and bureaucratization found in other organizations were absent from the RLF. Direct democracy—including even a provision for member referendums which had not been included in the constitutions of other interest organizations for many years—and high levels of member participation characterized the RLF (Elvander 1972, 83f.). The vitality of the union clubs in the 1980s is an important topic, as it has been claimed that the reorganization caused the union branch to lose much of its traditional democratic character and solidarity among members (Persson 1983; LRF Minutes 1972, Motion 4; speech by the chief executive officer (CEO) in LRF minutes 1974).

The regional level of the union movement is the 24 county unions (*läns- och provinsförbund*). The reorganization stipulated that members of their executive boards must come from both the union and cooperative branches of the farmers' movement, though local union clubs are guaranteed at least 50 percent of the seats. Their tasks include the representation of the interests of farmers on various policy issues which are relevant for the region, information on union and cooperative matters to the members, and certain responsibilities for the cooperative movement. It is the organizational unit which is commissioned to promote contact between the two branches within the LRF, a task which has proved to be a difficult one.

The county unions provide the central administration of the LRF with information on members and policy stands. Professionalism characterizes the level, as an increasing numbers of employed officials are responsible for the administrative tasks. An intermediate unit, the district union (*ortsförbund*), can exist between the local and regional level, in which case it coordinates the activities of the local union clubs within its jurisdiction (usually a municipality).

The cooperative movement is organized in a different way. The local units—the cooperative circles and districts (*kretsar, distrikt*)—belong to national associations (*branschorganisationer*). There is no direct relationship between the cooperative circles and districts and the LRF's central administration. In contrast to the union clubs, they are not regulated by the LRF constitution. Their size and tasks vary both within and between the national associations.

Cooperative societies (*ekonomiska föreningar*) are the regional unit. Two examples of a cooperative society are Arla, the largest one within the Swedish Dairies' Association and largest food producing enterprise in the Nordic countries, and Scanek, a meat marketing association that was created after a merger of four societies in western Sweden.

The current organization of the cooperative societies is professional and bureaucratic. There are 16 national associations. Before the reorganization, there were 13. With few exceptions, membership in the associations is based on the production of a particular commodity, for example, milk or meat. On an average, each farmer is organized in six associations. Table 3.2 offers information on the national associations. Also, it shows how farm rationalization and production specialization has affected their memberships, for membership rates have tended to decrease over the years.

Federalism characterizes the relationship between the cooperative associations and the LRF because they can decide their own tasks. The independent nature of the national associations is an important topic for debate within the LRF. Questions are raised regarding the character of cooperatives as business enterprises and their role in the countryside. Solidarity between the union and cooperative branches is another important issue in the 1980s. The trend is toward less independence for the associations, as explained in Chapter 6.

Reorganization rather than true amalgamation tends to characterize the organizational structure of the LRF central level, though true coordination is more apparent at this level. The union and cooperative branches have retained their own administration; legal technicalities explain this.

Responsibilities for the two administrations are held by the LRF Executive Board (*LRFs styrelse*). The Union Society (*LRF, ideell förening*) is

Table 3.2
Membership in the LRF National Associations[1], 1972—1988

	Date of Origin	Affiliation with SL	1972	1975	1980	1985	1988
KRF[2]	1963	1972			2,100	2,000	2,400
SAH	1861	1944	44,000	42,300	46,000	57,700	60,366
SBC	1889	1934—45	12,837	12,000	9,500	8,128	7,721
SBI	1907	1934	1,500	1,400	1,300	1,200	1,200
SF[3]	1914	1971		262,000	390,000	530,000	645,000
SFF	1930	1934	1,500	1,800	2,000	2,000	2,000
SJF[4]			216,000				
SHS	1945	1961	60,000	63,100	50,000	46,500	37,000
SKS[5]	1963	1972	3,200				
SLR	1905	1934	125,000	115,600	110,000	104,715	99,287
SMR	1932	1934	73,000	70,000	50,400	40,524	29,000
SOC[6]	1943	1944	7,900	12,500	18,400	22,740	25,548
SPOR[7]	1944	1980			7,700	7,100	5,700
SPR	1926	1937	1,000	800	609	638	647
SS	1933	1934	170,000	130,000	108,200	93,000	84,826
SSF	1927	1942	3,000	2,900	2,360	2,100	1,950
SSR	1932	1934	130,000	135,000	121,550	78,026	80,619
SÄ	1932	1934	15,000	7,500	2,575	1,288	1,064
Total number of members			863,937	856,900	922,694	958,135	1,084,328

Source: LRF, *Lantbruksåret* 1971, 1975, 1980, 1985, 1987; Johansson 1978.

1. The current national associations are: KRF, Association of Canning Crop Growers (*Konservväxtodlarnas Riksförbund*); SAH, General Mortage Bank of Sweden (*Sveriges Allmänna Hypoteksbank*); SBC, Swedish Sugar Beet Growers' Association (*Sveriges Betodlares Centralförening*); SBI, Swedish Distillers' Association (*Sveriges Bränneri-intressenter*); SF, Swedish Seed Growers' Association (*Sveriges Fröodlareförbund*); SFF, Federation of Swedish Cooperative Banks (*Sveriges Föreningsbankers Förbund*); SHS, Association for Swedish Livestock Breeding and Production (*Svensk Husdjursskötsel*); SLR, Swedish Farmers' Supply and Crop Marketing Association (*Svenska Lantmännens Riksförbund*); SMR, Swedish Dairies' Association, (*Svenska Mejeriernas Riksförening*); SOC, Swedish Oilcrop Growers' Association (*Sveriges Oljeväxtodlares Centralförening*); SPOR, Swedish National Association of Potato Growers (*Sveriges Potatisodlares Riksförbund*); SPR, Swedish Fur Breeders' Association (*Sveriges Pälsdjursuppfödares Riksförbund*); SS, Swedish Farmers' Meat Marketing Association (*Sveriges Slakteriförbund*); SSF, Swedish Starch Producers' Association (*Sveriges Stärkelseproducenters Förening*); SRR, Swedish Forest Owners' Association (*Sveriges Skogsägareföreningars Riksförbund*); SÄ, Swedish Egg Marketing Association (*Svenska Ägghandelsförbundet*).
2. From 1976 the new name for the SKS.
3. From 1975 the national association for the cooperative banks, the SJF is included in the SFF statistics.
4. Replaced by SFF.
5. Replaced by KRF.
6. Amalgamated with the SF in late 1988.
7. From 1980 a LRF member association.

a nonprofitmaking association that deals with union issues. A special council for confirming the proposals from the price-setting negotiations (*Avtalsrådet*) is included as an important subunit. The council is composed of 73 members elected by the general annual assembly. Members come from the county unions, and representatives of the cooperative societies must be included. The Delegation for Negotiation (*Förhandlingsdelegationen*) which negotiates commodity prices with the Consumers' Delegation is a unit within the Union Society. Nine elected officials comprise it.

The funds for conflict preparedness (*beredskapsfonder*) are under union jurisdiction. The constitution of the Union Society still includes a provision for referring issues directly to the membership in referendums. The provision has a more symbolic function, and is not used as a routine. The Cooperative Association (*LRF, förening upa*) is an economic association for dealing with matters of the national associations and businesses. The Farmers' Economic Company (*Lantbrukarnas ekonomi ab*) is a joint stock company which offers service to the member cooperatives.

As in most other large interest organizations, the Federation of Swedish Farmers is based on the principle of representative democracy. Members do not exercise direct influence on the regional or central organizational levels. They elect officials to the local levels who elect officials to the regional level who in turn elect 150 delegates to the general annual assembly (*Riksförbundsstämman*). The cooperative national associations elect 150 delegates to the assembly also. It is the highest decisionmaking body within the LRF.

Not only electoral democracy is filtered through different organizational levels. Expression of member opinion in the form of motions is dealt with in much the same way, a procedure which has met with some criticism and proposal for change (LRF minutes 1979, Motion 2). It is interesting that an overwhelming number of motions to the assembly come from the Union Society. It is very probable that the cooperative associations and their members view it as a union body. The cooperative movement may feel that its issues find a more suitable forum at the assemblies held by the national associations.

The general annual assembly discusses and decides on policy issues in other ways as well. One of its most important tasks is electing members to the two LRF executive boards. Half of the board members come from the union branch and half from the cooperatives. The two most important LRF leaders are the president, who is elected by the general annual assembly, and the chief executive officer, who is appointed by the LRF Executive Board.

The LRF central administration is in Stockholm, the capital of Sweden, a location which has occasionally led to membership criticism. Its organization is characteristic of all large Swedish interest organizations. It is pro-

Table 3.3
Personnel Employed at the LRF Central and Regional Administrations, 1971—1988

	1971		1975		1980		1985		1988	
Area	No.	%	No.	%	No.	%	No.	%	No.	%
Central	494	38	494	35	494	15	509	13	537	13
Regional	823	63	916	65	2,881	85	3,446	87	3,649	87
Total	1,317	100	1,410	100	3,375	100	3,955	100	4,186	100

Source: LRF, *Lantbruksåret* 1980, 1985, 1985; LRF, *Årsredovisning* 1976.

fessional, bureaucratic, and specialized. Tasks are dealt with by different departments. Until quite recently, the organizational structure had not changed much since 1971. Until 1989 the structure included a secretariat and separate departments for information, economic matters, personnel, price-setting negotiations, business and industrial policy, member recruitment and training, the cooperatives, taxes, consultation, legal issues, company law, the organizational newspapers, and the research institute.

Quite a number of people work for the LRF, as shown in Table 3.3. An important recruitment ground for the central administration has been the University of Agricultural Sciences (*Lantbruksuniversitetet*). In the past, most employees had farming as a characteristic of their social background. These days more people with urban backgrounds are employed. The regional and local levels of LRF have, when compared to the central administration, grown in staff size. The tendency is also present in other large interest organizations and reflects the changing nature of the Swedish state.

Currently the LRF owns entirely or together with its member associations or other organizations 17 joint stock companies. All are involved in serving the membership and representing their interests in one way or another. They perform the task of provision of protection or selective incentive that was discussed in Chapter 1. Examples include a health insurance company and clinic for farmers, a firm which administers help to farmers who need someone to substitute for them on their farms, a folk high school, and a publishing house. The latter two institutions are currently experiencing financial difficulty. The LRF has sold companies that did not serve farmers directly.

Ownership of joint stock companies is also common among the national associations. This form of ownership has proved beneficial for various economic and legal reasons. Yet the membership has not always viewed this ownership category as compatible with the principles of the cooperative movement.

Tasks

The LRF took over the tasks that were performed by the National Farmers' Union (RLF) and National Federation of Agricultural Cooperatives (SL). Recruitment of members to the farmers' union has remained an important activity. In particular, efforts are made to organize those who farm part-time. Active involvement of younger people is focused upon as the average age of members becomes older.

Many interest organizations include a youth organization; the LRF is no exception. Its youth organization (*Lantbrukarnas Ungdomskontakt*, UK) was created in 1971 and is active on the regional and national levels. A similar organization existed within the SL since the 1950s. The UK has close contacts with the other youth associations active in rural areas, including those belonging to the Center and Conservative Parties.

Training programs for members, active members, and employees are also given priority. The price-setting negotiations and contacts with government continue to be seen as fundamental.

Three basic tasks in the early 1970s were given higher priority. One is the political task of farmers' organizations discussed in Chapter 1. It reflects the more general changes in society and politics as well as the transformation of agriculture that put new issues on the LRF agenda. The general view is that political development in Sweden has a greater impact on all kinds of institutions. Organizations are forced to intensify their involvement in public policymaking.

For the LRF, the development signified that different policy areas were allocated more organizational resources. Depopulation of the countryside led the LRF to increase its involvement in regional policy. Forestry policy as well as the issues of property ownership and industrial and business policy were given more attention as a result of the structural transformation of agriculture. The aim of involvement in policymaking is to safeguard the interests of agriculture in state settings and to compete with other interests for political attention. Over the years a larger proportion of the LRF budget has been allocated to the political task. Now it accounts for 13 percent of the budget, compared to 5 percent in the late 1970s and 10 percent in 1982. The remaining LRF budget is proportioned as follows: 33 percent to the unit offering expert advice to individual farmers and cooperative associations, 12 percent to the LRF newspapers, 36 percent to the LRF-owned companies, 6 percent to other units (speeches by the CEO in LRF minutes 1974, 1982, 1988; LRF 1976).

Another function given more priority is public relations. It involves the tasks of collective bargaining, mythicizing, and politics that were taken up earlier. Informing the public about the farm problem is an issue which has increasingly demanded greater organizational attention and resources. The SL and RLF were the first Swedish interest organizations to experience how negative public opinion affected the ability of an organization to represent its members. The role that government plays in setting agricultural prices and formulating agricultural policy explains why the RLF considered the formation of public opinion to be one of its foremost tasks.

The first systematic efforts at opinion formation occurred in the 1950s. A joint SL —RLF committee was established in 1952 to inform consumers about the farm problem. Characteristic activities are campaigns that inform the general public on the farmer's share of food prices. Recent campaign material includes billboard and newspaper advertisements, stickers, video films, special folders, and handouts. Usually, the campaigns occur shortly before the yearly price-setting negotiations. The conflict fund of the union branch finances the campaign endeavors. The right of the union to call strikes has, in reality, been replaced by the use of public relations.

The importance given to opinion formation explains the more permanent contact with the mass media that was established in the 1950s. Presently, the LRF offers journalists special courses and study visits to farms and the cooperative asssociations. A final effort aims to increase contact between the agricultural sector, the farmers' organizations, and the public schools. Contact with the public school system is seen as a way to create a better understanding of the farm problem among the youth. Other large interest organizations have copied the methods developed by the organized farmers for contact with schools and the general public.

Public opinion formation and particularly information to consumers is a task that greatly concerns the LRF membership. Motions to the general annual assembly propose year after year that opinion formation be allocated more resources. The public debate on food prices in the early 1970s, as described in Chapter 4, made farmers keenly aware that information to consumers on the farm problem was a crucial task for the LRF. It was hoped that the new organizational newspaper would be able to play an important role here by informing consumers. Other efforts to reach consumers are taken up in Chapter 6.

The use of more traditional union tactics is a sensitive issue. Though the agreement on reorganization does not prohibit the use of delivery strikes as a conflict measure, it is generally understood that strikes are not feasible, as they would in the majority of cases be directed against the LRF cooperative branch. Words have become the unionized farmers' greatest tactic. Yet the economic situation facing most farmers may change this implicit organizational understanding. Dumping actions have occurred recently.

The third general task given priority after the reorganization was the LRF organization itself. Different concerns were attended to. The organization became self-critical. It was found that reorganization did not ameliorate problems. Rather they became intraorganizational.

Democracy within the LRF organization underwent appraisal in light of the difficult situation that all popular movements experienced in the late 1960s and 1970s. Solidarity within the membership tended to decline. Members began to perceive that the organization did not represent a common interest and that they differed from other members whose farm size or production was dissimilar. Bureaucratization and member alienation were discussed more openly.

Heterogenization of the membership is a factual development as well as an issue of member attitude. Guilds were created inside the LRF, within which members with similar agricultural production believe that they share both a common interest and sense of solidarity. Currently, many of the guilds function as internal lobby groups. Protest groups which either criticized general or particular LRF policies also began to organize in the early 1980s.

Also, coordination of the cooperative and union branches needed attention, and solutions were sought to the problems. A committee was commissioned by the general annual assembly in 1977 to study the LRF's position and function within the farmers' movement. Its report touched upon many of the following problems (LRF 1979a).

A key issue involves the discrepancy in membership rates in the union and cooperative branches. Membership in them is based on different logics of collective action. For the union, a large number of members is traditionally seen as leading to more influence. A high organizational density lends legitimacy to the interest organization. There is said to be power in numbers. For the cooperatives, an individual farmer calculates how beneficial membership will prove to be in economic terms. Whereas the spirit of solidarity is an important reason for union membership, individual economic utility is the principle for membership in the cooperatives. The important questions for the LRF here are whether union democracy and business efficiency can be successfully combined and whether individual farmers can be convinced that joint membership is essential.

The cooperative enterprises is another main area targeted for improvement. Different problems are addressed. Increased competition from other food processing companies means that marketing must be given higher priority. Economic considerations, such as those regarding investments and capital, is a second problem. Democracy and member activity within the economic associations is yet another issue. The actions taken by state agencies against the business practices of the cooperatives were also addressed. Effort in the 1970s was made to formulate policy norms for these as well as other issues of importance for the future of the farmers' cooperative movement (LRF 1979b). The essential issue is whether the cooperatives can survive according to the principles of cooperation in the increasingly complicated and competitive economic and political climate of the late twentieth century.

4

Exceptionalism Questioned

CRITICISM OF AGRICULTURAL POLICY

Dissatisfaction with both the goals and the results of agricultural policy increased as Sweden became a postindustrial nation. The stands taken by the different actors involved in policymaking and even their lines of reasoning reflected not only old concerns but newer ones. Four important issues of the late 1950s were farmers' income, food prices, farm rationalization, and a solution to surplus production. While the concerns were important in the 1960s, other issues were emphasized as well. They included efficiency, centralization, the environment, and consumer interests, all of which are characteristic of the time period.

The goals established in the 1947 parliamentary decision underwent modification in the other main decisions on agricultural policy. The second major one was made in 1967. As in 1947, it was preceded by an investigation that was conducted by a royal commission. The commission was appointed in 1960 for different reasons.

A number of the predictions on self-sufficiency and farm rationalizations made by the 1942 royal commission were wrong. While self-sufficiency in food production was reached in the early 1950s rather than in the mid-1960s as predicted by the commission, external farm rationalization went slower than forecasted.

General planning was also considered to be unsatisfactory. On the one hand, coordination of policy implementation was impeded since the policy tasks were divided among separate government units. On the other hand, some of the 1947 policy aims on farm rationalization were found to lack adequate definition for their successful implementation.

The role of agriculture in the national economy is the third reason. The setting of agricultural prices according to the 1947 guidelines was seen as a policy failure because it neither sufficiently promoted rationalization through stability and agricultural specialization nor income equality. There was also a growing concern about the economic exceptionalism of agriculture. Two of the main questions that were raised regarded the status of agriculture in comparison to other businesses and industries and the allocation of public finances to agriculture in relation to other groups in society (Julen 1982; Torpe 1982; SOU 1966:30, 1966: 31; GB 1965, no. 100).

The reasons for policy reevaluation reflect the spirit of the times and the social democratic evaluation of social progress in these years. Centralization and planning were seen as solving most problems. A national economic perspective was considered necessary to reach decisions on the basis of distributional policy for it faciliated the comparison and evaluation of the situation of different groups.

Another important reason for the decision to appoint a royal commission reflected the time period as well. It involved the competition among different organized interests for participation in policymaking and the decision to incorporate interest organizations in the government process. Countervailing powers and corporatist integration are both important in this period, as discussed below. Also, the Riksdag pressed for a new investigation for it felt that the public administration and organized interests were too influential in the agricultural sector (Odhner 1988).

In addition, most Swedish farmers were not satisfied with their situation in the late 1950s and 1960s. They were fewer in number and felt that they had lost political support and influence. As discussed in the preceeding chapters, the Agrarian Party desired to identify itself with other groups in society. The Social Democratic Party was seen as less willing to give consideration to the interests of farmers and the minister of agriculture as more concerned with consumers than producers.

In fact, the group of social democrats that was more sympathetic to the situation of small farmers in particular were losing influence within the party. It became easier for the LO to approach the government directly and criticize food prices and agricultural policy after the agrarians left the coalition. An expansion of agricultural policy was going to take place.

In 1958, the minister proposed changes in agricultural policy. The farmers' organizations were sounded out on the new orientation in the 1959 negotiation on agricultural prices, a common forum for the definition of policy issues and refinement of policy goals. One of the most controversial and important decisions reached in the negotiation was the calculation of commodity prices on the basis of a larger farm (*normjordbruket*) of 20 to 30 hectares rather than the basic farm (*basjordbruket*) of 10 to 20 hectares. The issue of farm size was debated at the Social Democratic Party congress in 1960.

The decision is interesting in two ways. First, over 85 percent of all Swedish farms at that time were smaller than 20 hectares, and over 60 percent were less than 10 (Anderson 1987, Table 2). The decision meant that price calculations would either force small farmers out of business or force them to rationalize further, a line of reasoning quite similar to the policy of the workers' movement on rationalization of industry through the wage policy of solidarity.

Second, the two largest farmers' organizations accepted the larger farm size as the norm. While the new principle did not meet with disagreement within the National Federation of Agricultural Cooperatives (SL), the National Farmers' Union (RLF) experienced serious internal conflict—49.45 percent of the delegates at an extra congress were against and 50.55 percent were for the principle—since most of its members had smaller farms (Torpe 1982). The organization would support a more radical policy of farm rationalization if certain conditions were met.

The government presented its bill, which even cited the RLF's decision and its conditions for support, to the Riksdag in 1959 (GB 1959, no. 147). Other bills, motions, and statements on the agricultural issue were submitted as well. The Riksdag voted on the agreement that was made on agricultural prices but decided to await an investigation before it voted on the issue of the criterion of farm size for price-setting. The issue was controversial, and the Standing Committee on Agriculture proposed that a royal commission be appointed to investigate the matter fully.

The debate in the Lower Chamber offers important insight into the thinking of the time and the dilemma facing both the Center Party and the farmers' organizations. Different MPs accounted for the discussions on the subject within the two interest organizations for farmers, the RLF and SL. The RLF was considered to be a decisive actor on the issue of farm size and to have influenced the Center Party greatly. The close connection between the Center Party and the two organizations was brought up in the parliamentary debate. It was questioned whether the Center MPs (including the party leader Gunnar Hedlund) who sat on the executive board of the SL represented the Center Party or another interest in the parliamentary debate. The ties between the party and the farmers' organizations are still a rather common political criticism of the Center Party.

That the farmers were forced into a trade-off was emphasized also. A minister of agriculture from the Agrarian Party in the coalition government of the 1950s entered the debate to state that he basically supported the government bill. Other Center MPs were more skeptical and stressed the importance of the smallest farms for war preparedness, politics, and the economy of the countryside. As pointed out by at least one speaker in the Riksdag, peasant romanticism played a part in the argumentation of the Center Party.

The parliamentary situation can only be described as very difficult for the Center Party. While in its motion to the Riksdag it criticized the proposal for a new farm size for price-setting, it was much more reserved in its criticism in the statement by its members to the Standing Committee on Agriculture. Some politicians expressed their dislike of this double-edged message. Yet division in the committee would have jeopardized broad political agreement on the main issues of agricultural policy, thus violating an important tenet of Swedish political culture in these years.

Serious criticism of the proposal for a new farm size for price-setting by the Center Party in its statement to the committee would have signified that the party, in reality, vetoed the agreement made by the farmers in the price negotiation. Yet the contradictory messages of the Center Party's parliamentary motion and committee statement reflect disagreement among the party's MPs on this issue as well.

A majority of the Riksdag in the Upper and Lower Chambers approved the government bill that signaled a new orientation in agricultural policy. The vote was not easy for the Center Party: 30 out of 32 MPs in the Lower Chamber voted against it, one abstained, and one—the leader of the party—was absent for the vote (Riksdag minutes (RP) Lower Chamber (AK), 1959, no. 13, 20, 21; Torpe 1982; Julen 1982). Perhaps his absence reflected how sensitive the vote was for the party. The party voted against the bill in the Upper Chamber. The other non-socialist parties approved it in both chambers.

While agricultural policy was debated politically, farmers experienced economic problems. Farming had become a capital-intensive operation. Extensive investments were necessary for farmers to keep up with the technology that was needed for rational production. Technological progress meant that fewer farmers were needed for self-sufficiency, and many were leaving the countryside for employment in industry. Farm income was also lower than prescribed by the income goal of Swedish agricultural policy.

Agriculture had become less important for Swedish economic progress. It dropped from 5.9 percent of the gross national product in 1950 to 3.5 percent in 1960. The trend continued in later years: 2.1 percent in 1970 and 1.9 in 1982 (SOU 1984:86, Table 2.2). Farmers were increasingly becoming specialized producers, and fewer farmers were needed in general, as a result of the policy of rationalization. All changes affected the activity of the farmers' organizations.

A royal commission was appointed in 1960. The government's instructions stated that its task was to study how much agricultural production Sweden needed in an economic, war preparedness, demographic, and social perspective. Also, the effects of structural changes in agriculture were to be considered.

The composition of the committee differed from the 1942 one and reflects the time period. Civil servants and representatives from interest organizations were more common among the committee members. The RLF and SL each had one seat, as did their joint research institute (*Jordbrukets Utredningsinstitut*, JUI, later renamed *Lantbrukets Utredningsinstitut*, LUI). The consumers were represented through the KF and two LO unions. The farmers' organizations were well-represented among the commission experts as well.

Before the final reports of the committee were published in 1966, the relationship between the farmers' organizations and the social democratic government deteriorated greatly. The reason was a proposal for a new law on land use, which the farmers saw as leading to the death of the countryside. The RLF newspaper included a number of articles on the proposal. One was entitled "Your Village Doomed to Die?" (*Din bygd dödsdömd?*) (*RLF-tidningen* 1968, no. 14). An advertisement campaign commenced also. One read "Cultural District Today—Wasteland Tomorrow" (*Kulturbygd idag—ödebygd imorgon*), a second proclaimed "Cultivated Land, Culture Landscape, Recreation Land" (*Odlingslandskap, kulturlandskap, fritidslandskap*). It is clear that the aim of the campaign was to capture the attention of urban residents by using the rural myth.

The climate was not the most appropriate for a peaceful discussion of the proposals included in the committee report. Its proposals differed from the 1947 agricultural decision in important ways. Agricultural effectivity and efficiency, as measured in terms of ongoing technical and economic developments, were highlighted as important goals.

Unlike the past, state measures were not considered to be social support for farmers. Rather, the commission stressed that state intervention should facilitate and accelerate the transfer of labor and capital to more productive sectors. It proposed the abolishment of the smallholders' subsidy that was introduced in 1947. Also, agricultural prices were not to increase as quickly as the general price level until rationalization and reorganization proved effective. It was further proposed that self-sufficiency in peacetime could decrease without jeopardizing Swedish war preparedness. (Years later it was pointed out that the embargoes of Sweden in the first and second world wars were not complete and that the level of war preparedness could be lowered without endangerment.) Other goals were included in the commission's report but not a policy goal for income equality (SOU 1966:30, 1966: 31; Thullberg 1980).

Not all committee members agreed with the proposal. Eight reservations as well as 19 special statements from commission members accompanied the final report. The lines of cleavage on agricultural policy were deeper than they had been in an official setting of this kind in the past. There were two opposing groups. One included the KF, the Social Democratic Party, and the trade unions and articulated the interests of consumers in agricultural policy. The smaller group was composed of the representatives of the farmers' organizations and the Center Party. It articulated the interests of the farmers.

The representatives of the consumers thought that the commission had given undue attention to the issue of the countryside and demographic changes. They felt that agricultural policy should be evaluated in economic terms only. Yet both groups agreed that the main object of agricultural policy was to help agriculture adapt to economic and technical changes.

The issue of controversy was the rate of the change proposed in the report. The farmers believed that it was too rapid, an argument reminiscent of the very early debates on agriculture. They raised the issue of the effects of change on the countryside and challenged the commission's calculation of Sweden's rate of self-sufficiency in peacetime. War preparedness was becoming an issue of disagreement for the contending interests as well as a symbol in debates on agricultural policy.

Another explosive issue was preservation of the countryside and Swedish landscape. The representatives of the consumers were forceful in their view that farmers should not become caretakers of the environment. That arable land would, as a consequence, turn to meadow and forests was not seen by them as a sufficient reason for public support to farmers. They stressed further that agricultural policy was not the correct policy forum for the issue of tourism and recreation. The issue of landscape preservation was dominant in the public debate at the time. It is mentioned here because the issue is salient in the current debate in Sweden as well as in other countries (Drake 1985, 1987; Horsfield 1986; EC 1986, 1987).

In 1967, the government submitted its bill to the Riksdag. Generally, it supported the commission's proposals on all important points. Agricultural policy was to change dramatically. Exceptionalism was to end. The minister wrote: "This means that agricultural policy will, more than is presently the case, have the same main purpose as general economic policy....The objective of agriculture should, as far as possible, be the same as those for business and industry...." (GB 1967, no. 95, my translation). He further stated that food prices should be kept as low as reasonably possible and that the main reason for the maintenance of a large domestic agricultural production was good war preparedness. He stressed that agricultural policy should not promote landscape preservation.

A majority in the Riksdag can be said to have accepted the new guidelines as formulated in the bill. The social democrats supported the proposal. The Center and Conservative Parties opposed it in both chambers. The liberals abstained. This was a tactical move explained by the importance that agricultural policy played for cooperation between the Liberal and Center Parties in the 1960s. The policy area is difficult for the party because liberal voters are mainly urban residents. The communists abstained in the Lower Chamber. The representatives of the farmers' interests were not very influential on the issues of the size of production and principles for price-setting.

The political situation confronting the farmers' organizations in the 1960s can be described in the following way. It seemed that agricultural realism would replace peasant romanticism as it was decided that farming did not have intrinsic value. Farm life was not to be seen as building better national character than city life, and life in the cities was judged to be good. Family farming was seen as an inefficient means of production. Agriculture was still associated with nationalism, for it was an aspect of war preparedness. Attempts to use other than economic values in analyzing the role of Swedish agriculture in the postindustrial era were criticized decisively. The value of the countryside for scenic reasons was simultaneously acknowledged and disregarded as an issue related to agricultural policy. The parliamentary decision marks the first expansion of the parameters of agricultural policy.

ORGANIZED CONSUMER DEMANDS ON AGRICULTURE

One of the main reasons for the decline of the political influence of agriculture is, as discussed in Chapter 1, the existence of a countervailing power. The most important countervailing power in agricultural policymaking is the organized interests of consumers. A consumer movement, as in the United States, does not exist in Sweden. Rather consumers are represented through established interest organizations, political parties, in particular the Social Democratic Party, and the state. The influence of consumers began to grow in the 1950s.

That the state through its institutions safeguards the interests of consumers has been singled out as a characteristic of Sweden (Schonfield 1965, 205, 390). Public bodies which safeguard the rights of consumers began to develop after the war. Two examples are the the National Price and Cartel Office (*Statens Pris- och Kartellnämnd,* SPK) from 1957 and the Office of the Competition Ombudsman (*Näringsfrihetsombudsmannen*, NO) from 1954. Both are highly influential in agricultural policy. After 1957, consumers have been represented on state bodies through the trade unions and the Cooperative Union and Wholesale Society (KF).

The trade unions and consumer movement decided to attempt to influence agricultural policy in a more systematic and organized fashion in 1953. That year the Consumer Committee for Agricultural Policy (*Konsumentkommittén för jordbruksfrågor*) was created by leaders of the KF, LO, and TCO. The smaller peak union organization, the Swedish Confederation of Professional Associations (SACO), did not participate.

The associations felt that the position of the farmers' organizations had been strengthened to an unreasonable degree as a result of the 1951 coalition government. They wanted to counter the influence of farmers on agricultural policy. They demanded that the interests of consumers be given seats on the county agricultural boards which were dominated by representatives of the farmers' interest. Their request was granted in 1967.

The committee emphasized that the opinions it expressed were its own, but it is reasonable to assume that they were influential within the three organizations themselves. Two of its members even represented their respective organizations on the 1960 royal commission. The committee published four books which generally questioned agricultural exceptionalism and argued for more consumer-oriented policymaking.

Though the committee was abolished in the mid-1960s, the organizations continue to represent and articulate the interests of consumers through their representation in the Consumer Delegation (*Jordbruksnämndens Konsumentdelegation*, KD) that is attached to the National Agricultural Market Board, in remiss replies, participation on committees and groups as well as through official statements.

ROLE OF COUNTERVAILING POWER IN POLICYMAKING

Gradually, agricultural policy began seriously to interest other groups in society than the farmers. Organizations representing consumers became active. The agricultural subgovernment became more accessible to other groups. Formal recognition by the government that other interests should directly participate in the agricultural sector came in the 1960s. A transition from an iron triangle that primarily involved the ministry of agriculture, the standing committee in the Riksdag, and the farmers' organizations to an issue network of agriculture, which included other societal interests, occurred at this time.

Representatives of the consumers became legal participants in agricultural policymaking in two direct ways. The Riksdag decided that consumers should be given direct influence over regulatory matters through representation on the executive boards of the seven commodity regulatory boards. Initially, the KF was their only representative, but the government later appointed representatives from the LO, SACO/SR (*Centralorganisationen SACO/SR*), and TCO to represent the interests of the public.

An evaluation conducted by the Office of the Parliamentary Auditors (*Riksdagens Revisorers Kansli*, RRK) found that the representatives of the consumers had difficulty in coordinating their actions and lacked a sufficient understanding of the rules that applied to price negotiations. One problem was that statistics on the market situation for commodities, which form the basis for decisions in the regulatory bodies, were compiled by the farmers' organizations (RRK 1979). Consumer influence is still lower than the producers, and their representatives have been characterized as in a hostage situation (interview with Nyberg).

The main way that consumers became a countervailing power in agricultural policymaking is through the Consumers' Delegation created in 1963. The government reasoned that it was important for international and domestic reasons to give consumers the formal opportunity to articulate their interest in agricultural policymaking.

Implicitly, the minister acknowledged that agricultural policy was undergoing change and expanding. It was becoming an issue area concerning consumers as much if not more than farmers. Trade in agricultural commodities had begun to increase, and it was considered necessary for consumers to form an opinion on the importing of food to Sweden. Here the basic issue is the price of food.

A council (*Statens Jordbruksnämnds Råd*) associated with the National Agricultural Marketing Board since its creation did not, due to the development of the system of price negotiations, perform the role invisioned for it. The government desired to underscore the importance of nonfarmer influence on agricultural policymaking by creating a new body, the Consumers' Delegation.

In an interview long after its establishment, the minister of agriculture at the time explained that he wanted to create a countervailing power by political means in order to check the domination of the producers in the agricultural and food sector (Anderson 1987, 45ff.). The government stated in its bill that two of the members of the delegation should be women, a decision which can be understood as a continuance of policy from the past as home economic associations were involved in previous public bodies (GB 1962, no. 193). The prototype for the organization of the new institution was the Farmers' Delegation for Negotiation (*Jordbrukets Förhandlingsdelegation*).

Neither the farmers' organizations nor the non-socialist parties were pleased with the proposal, which had been sent out on a remiss referral. The SL and RLF submitted a joint reply in which they maintained that no study showed that the council attached to the Agricultural Marketing Board did not allow for sufficient representation of other social interests and that the report from the ministry had underestimated consumer influence. Yet the two interest organizations of farmers notified the ministry that they would not oppose the creation of the Consumers' Delegation (GB 1962, no. 193).

The RLF was less restrained in its newsletter to its local leaders. It criticized the ministry for not sounding out the farmers' organizations before the proposal was sent out on remiss referral and announced that it opposed the creation of the delegation in its remiss reply (*RLF-information* 1962, nos. 6, 8, 10).

A motion opposing the reasoning in the government bill was submitted by MPs from the Center Party to the Lower and Upper Chambers of the Riksdag. They criticized the government for focusing its attention on the price negotiations, stated that higher food prices were caused by middlemen, and proposed that the farmers' cooperative movement should be represented on the Consumers' Delegation. They emphasized strongly that consumers had more influence over agricultural policy than farmers (RP 1962, AK no. 925; RP 1962, FK no. 764).

A majority of the Riksdag's Standing Committee on Agriculture supported the proposal but described the main purpose of Consumers' Delegation as a channel for the distribution of information on the complexity of the farm problem (JoU 1962, no. 34). The Riksdag approved the bill. Bloc voting characterizes the result (RP 1962, FK no. 35; RP 1962, AK no. 36).

The Consumers' Delegation grew in importance over the years. Its organization was a topic of discussion and evaluation in the late 1960s and 1970s. The influence that it exerted in the price negotiations was considered to be satisfactory. In fact, it compared favorably to the Farmers' Delegation for Negotiation in status and influence. Officials employed at the Delegation's office stated in an interview that they strongly believed that the influence of consumers had increased at the expense of the influence of farmers (RRK 1979; GB 1969, nr. 107, 1975—76, no. 154; DsJo 1976: 1; Riksdag motion (RM) 1975—76, no. 1272).

The position of the Delegation was strengthened further in 1976 when the

Riksdag voted to give it additional administrative resources (GB 1975—76, no. 154; RP 1975—76, no. 127). A more recent assessment shows that the Delegation functions as a check on more adamant demands that could be made by the farmers. Yet it should be emphasized that its main task is not to negotiate as low food prices as possible. Its participation in the price-setting negotiations is subject to the parameters of Swedish agricultural regulation, which has not had low food prices or market competition as its point of departure (interview with Nyberg).

Representation on the Delegation it still discussed. The main issue is how the interests of the consumers should be defined. Some actors believe that it is important that only organizations that are specific consumers' organizations retain representation because others, such as the unions, must balance different interests and compromise the special interests of consumers. Another line of reasoning is that consumers are those who buy farm products, that is, retailers and wage earners.

Representatives of the consumers have also become more knowledgeable and adept in the agricultural subgovernment. The LO and TCO are attempting to create a stronger profile as safeguarders of consumer interests (TCO 1978, 1981, 1987; LO 1985). The new chairman of the Consumers' Delegation is the TCO's expert on agricultural policy. Former chairmen have been civil servants of high rank, which has hindered them somewhat from debating the agricultural issue publicly. Now the Delegation is led by someone who discusses the issues in the mass media, a characteristic which has been said to be an important reason for his selection by the minister of agriculture (interview with Nyberg).

The political debate on the role of consumers in agricultural policy is a good example of classical Swedish political thinking. All the concerned interests in a particular policy sector participate in the decisions. The interests are represented through established interest organizations, which are sounded out on official and informal proposals. In this way, decisions are legitimized and anchored in the memberships of these organizations.

Controversy can arise over the proper balance of the representation of interests. Yet it is extremely difficult to argue politically against representation for all concerned interests. Gradually, the interests become countervailing powers. The only potentially convincing argument against this Swedish concept of fairness is a different evaluation of the balance of interests. The Center Party and the farmers' organizations evaluate the balance differently and use it to attempt to limit the influence of the consumers over agricultural policy.

POLITICAL PARTIES AND FARMERS

Swedish political parties have always competed for the farmer vote. As discussed in Chapter 2, the attention which the parties give to the farm problem can partly be explained by a desire to attain more votes. Until the 1960s, this occupational group was the only group that was the object of such electoral attention. While other groups voted on the basis of class, the farmers voted after their interests, which can be represented by different political parties. Since that time, another group, the white-collar workers, has grown in number and received the same kind of political attention previously shown to the farmers. Yet their importance does not mean that the farmers are neglected in electoral politics.

The farm vote is still very important for the Center Party. Attempts by the party since the 1950s to broaden its electoral base have not been completely successful. One explanation is that farmers are reliable both as voters and party workers. In the 1985 general election, most farmers (64 percent) voted for the Center Party, and the occupation is also the party's largest electoral group (26 percent). The farmers' movement is important in electoral campaigns and in the party organization. Farmers have a high and reliable electoral participation: 90 percent cast votes in 1985, with the average for the entire electorate at about the same level (Holmberg and Gilljam 1987, 74, 179).

A major problem for the Center Party is that farmers comprise only 4 percent of the electorate. Yet as the balance between the two political blocs is very even, all votes are important. Another problem for the party involves the discrepancy between environmental and agricultural policy, the two cornerstones of its political profile.

The environmental issue has become extremely important in Swedish politics. In the 1970s, the Center Party was identified as an environmental party, a factor which helps explain its electoral success in this decade. Concern for the environment drew voters from groups which did not traditionally vote for it.

The issue is problematic for the party. While the newer groups of voters criticize not only nuclear power but also the use of chemical pesticides in agriculture, the farmers have up until recently been unwilling to discuss the use of agricultural chemicals. In 1985, 78 percent of the electorate favored a ban on chemical pesticides (Holmberg and Gilljam 1987, 125).

The cleavage within the party is serious since many parties identify themselves with the environmental issue now. It was the social democratic government that established the newest ministry, the Ministry of Environment and Energy (*Miljö- och Energidepartementet*), in 1985. Other issues, such as the centralization of the farmers' cooperative movement, also divided the Center Party's old and new guard and are emphasized by other political actors.

Attempts by the party to keep and capture voters have led to a reevaluation of the relationship of the party to farmers and their organizations. This is necessary because the party does not want to lose more votes. Votes were lost in 4 consecutive parliamentary elections: from 25 percent of the general vote in 1973 to 12 percent in 1985 (Holmberg and Gilljam 1987, 23). The trend was broken in the 1988 election. The present discussion within the party emphasizes the relationship between the party and the farmers' movement and the need to formulate an environmental agricultural policy.

A second reason for the continued importance of farmers and agriculture in politics is concern for the countryside. While the Center Party is most reliant on the vote (with about 45 percent of its votes being rural in the 1985 election), other parties rely somewhat on support from strictly rural districts. Eleven percent of the votes cast for the Conservative and Liberal Parties and about 10 percent of all social democratic votes are rural (Holmberg and Gilljam 1987, 179, 199ff.). The Social Democratic Party lost almost half of its support among farmers in the 1960 election and has never retained them (Isberg et al. 1974).

The even division between the two political blocs makes strictly rural areas electorally attractive for other parties as well. The large cities are rapidly growing in size and are increasingly voting non-socialist. The social democratic stronghold is the smaller towns and cities, two geographic areas that are losing inhabitants to the large cities.

Electoral research has further shown that rural and urban inhabitants—regardless of occupation and class—tend to cast their votes differently. Whereas class voting is declining, urban/rural voting differences are increasing, particularly among younger voters in the 1980s (Holmberg and Gilljam 1987, 222, 312).

Also, history has shown that disregard of the issue may cost electoral votes. In 1963, the Social Democratic Party began to revise its position on agricultural policy. A revised platform was presented in 1966. Many people considered it to be very consumer-oriented, and the farmers' organizations criticized it adamantly. The low election return for the party in the 1966 municipal election has been attributed to this (Isberg et al. 1974). An investigation conducted by the party shortly after the election showed that other reasons explained the bad results. Afterward, the party reevaluated its platform and decided to formulate a more consistent business policy that also included agriculture.

The final important reason for the importance of agriculture in party politics today is the role that agricultural policy and farmers play in the attempts at interparty coordination. The Center Party is crucial for three different efforts in coalition building. The other parties have tended to play down their demands on agricultural policy to insure cooperation from the Center Party.

The classical government coalition is the social democratic/agrarian one. It cannot be ascertained whether the social democrats consider this alternative a desirable one today, as scholarly study of the subject is lacking. It can be argued that social democratic reaction to certain political events reflects an attempt by the party to attend to the interests of the Center Party. It has been noted that they did not criticize the Center Party harshly in the 1988 electoral campaign. Yet it does not seem that this has the same significance for the interests of farmers today.

The close relationship between the Center Party and the farmers has definitely played an important role in negotiations with the other two non-socialist parties. The relationship affected greatly the unique cooperation (*mittensamverkan*) between the Center and the Liberal Parties in the 1960s. The official cooperation benefited the Center Party and its main constituency, the farmers, more than the Liberal Party and its urban voters. Sharp criticism from the liberals on agricultural exceptionalism was not formulated in these years. Also, it aided the Center Party in its attempts to modify its agrarian identity (Möller 1986).

Agricultural policy played a role in non-socialist party cooperation in the 1970s as well. That the Center Party gave priority to the interests of farmers in the non-socialist coalition government created after the 1976 election is clear. The party convinced the other parties to appoint a leading center politician as minister of agriculture; more money was appropriated to the ministry (see Table 4.1), and the statement from the government on its agricultural policy took other than economic concerns into consideration. An enhanced position of influence for the interests of farmers was also perceived by the leaders of the LRF, who were satisfied in general with the situation (Hammerich 1977, 276; Verba et al. 1987, Figure 8.6).

The urban and consumer-oriented Liberal Party did not always agree with the Center Party on agricultural policy in the years in government. Leading liberals still criticize the Center Party for catering to the interests of the farmers. The farmers' movement is also a focus of criticism (Bergström 1987, 218, 341; Samuelsson 1987, 113—18; RM 1986—87, no. K212; speech by Westerberg 1984). Yet liberal criticism of the very influential role of the organized agricultural interest in politics does not prohibit the party leader from cultivating an agricultural image, for example by posing in a barn next to a few cows for a picture in the largest Swedish daily newspaper

(*Dagens Nyheter* 4 June 1985). In sum, farmers have benefited from the attempts by both the Social Democratic Party and the non-socialist parties to cooperate with the Agrarian/Center Party in elections and coalitions.

AGRICULTURAL POLICY IN POSTINDUSTRIAL SWEDEN

Debate on agricultural policy in the early 1970s focused on the decline of the countryside and food prices, which rose 29 percent from 1969 to 1972 (*När, var, hur* 1973). The mass media gave attention to a group of women in a Stockholm suburb (*Skärholmsfruarna*) who organized a protest in 1972 against high food prices. They boycotted meat and milk products for one week. Other people joined their protest. A demonstration—the first of its kind since World War I—drew thousands of people in Stockholm, and activities took place in other parts of Sweden as well. The political parties took the outburst of militant consumerism seriously. The Riksdag debated the price issue, and the government revised its position on food subsidies.

Price stops were first introduced for staple foods and later replaced by subsidies for meat and milk products. The subsidies were an economic steering measure aiming to satisfy consumers and attempting to solve problems that were not agricultural in nature. When generally abandoned in 1984, they created surpluses of milk and meat and problems for farmers and their organizations. It is not common knowledge that the surpluses were a consequence of the consumer demand created by the subsidies, though policy documents agree on this fact.

The 1970s can convincingly be called the decade for the interests of consumers in Sweden. Consumer protection was institutionalized. Legislation was passed. Public bodies with a consumer orientation were established. They are the Market Court (*Marknadsdomstolen,* MD), Consumer Ombudsman (*Konsumentombudsmannen,* KO), and National Board for Consumer Policies (*Konsumentverket*). Many of the institutions have voiced serious criticism against the farmers' cooperatives. The ministry was also affected, in that the National Food Administration (*Statens Livsmedelsverk*) was placed in its jurisdiction. Other changes that signified a reorientation of the ministry toward consumerism and environmentalism came later.

At the same time, farmers experienced a more favorable attitude to agriculture from policymakers. A third important commission on agricultural policy was appointed in 1972 by the social democrats. Its final report was presented to the non-socialist coalition government in 1977 and offered a different interpretation of the goals of agricultural policy. Economic considerations, farm rationalization, and war preparedness were still considered important, but other concerns—such as regional and environmental policy as well as the income of farmers and their standard of living, food of high quality and low price, and preservation of the countryside—were given priority (SOU 1977:17). The bill presented by the government followed the guidelines proposed by the commission (GB 1977—78, no. 19). A second expansion of the parameters of agricultural policy would take place soon.

There are different reasons for the reorientation. The international situation had changed how developed countries viewed the problem of surplus production, as they decided to help solve the problem of starvation in the world. This viewpoint was expressed in discussions in the GATT in the late 1960s. Food products were exported to third world countries. The bill underscored the importance of high production for this reason. Also it was predicted that world food prices would be sufficiently high to absorb the domestic costs of surplus production. Finally, Swedish industry did not need more manpower in these years; a central argument for farm rationalization was no longer valid.

Rapid farm rationalization was seen as leading to rural decay, a state of affairs that the non-socialists declared that they wanted to improve while in office. They emphasized the role of agriculture in an active regional policy. Public concern for the environment was growing also. Environmentalists criticized the use of agricultural chemicals, and the public became aware of the role that farmers could play in the preservation of the countryside landscape. A new and important aspect of the reorientation was the recognition of part-time farmers and the decision to give them public support. The political composition of the government and Riksdag is significant for the orientation because the non-socialists were in the majority and could push their policy stand through Parliament.

It is worthwhile to digress in order to offer an account of the social democratic reaction to the government bill. The party emphasized that it believed in broad political unity on the issue. What it meant was that the policy should not be approved without its support. The party voted against the general policy guidelines and a number of the goals formulated in the government bill.

Yet it is intriguing to note that a state of broad agreement, when defined as general agreement among most of the political parties of the Riksdag, had not existed for decades. In the 1970s, the social democrats chose clearly to safeguard the interests of consumers over those of producers in its stands on agricultural and rural policy (JoU 1977—78, no. 10; RP 1977—78, no. 54; Elander 1978, 96ff.).

The agricultural policy promoted the interests of agriculture in a new way. One author calls this period the golden years for Swedish agriculture

(Anderson 1987). Exceptionalism continued to characterize the Swedish perspective on agriculture.

FROM AGRICULTURAL TO FOOD POLICY

A new investigation on agricultural policy was commissioned once the social democrats returned to office in 1982. The instructions that the minister of agriculture gave to the commission show that agricultural exceptionalism was challenged again. Its name was the Commission on Food Policy (*Livsmedelskommittén*); its task was to study the situation of Swedish agriculture in a broader perspective that evaluated the traditional and basic three goals of agricultural policy, the consequences of surplus production for government and consumers, the food processing industry, and the goals specified in the 1977 Riksdag decision.

As with most commissions appointed by the new government, the commission worked quickly. A preliminary report was finished in 1983 (DsJo 1983:10). A bill based on the report was debated in the Riksdag in 1984, whose decision signified that war preparedness should be the main goal of agricultural policy again (GB 1983—84, no. 76; RP 1984—85, no. 77). The decision paved the way for the fourth major agricultural policy decision in Sweden. The commission presented its final report in 1984. It emphasized that agriculture is only one aspect of food policy (SOU 1984:86).

One change that influenced the Swedish debate on agricultural policy is the issue of deregulation. Swedish economists have repeatedly stressed that agricultural protection should cease for economic reasons. A book published in 1984 provoked a debate on the issue by arguing adamantly for deregulation of Swedish agriculture (Bolin, Meyerson, and Ståhl 1986).

By far the largest proportion of budgetary allocations for the agricultural sector goes to subsidies. It is considered a serious problem, though public financing of agriculture is only a small fraction of the national budget, as shown in Table 4.1. The commission viewed deregulation as a good idea but stated that it was impossible for Sweden to deregulate before other countries.

A second issue influencing the work of the commission was food prices. Different aspects of the problem are important here, as consumers and producers are both hit hard by high food prices. Consumer prices on food staples have increased dramatically when compared to other commodities. High food prices tend to influence other areas of the Swedish economy, such as collective bargaining, inflation, and the standard of living.

The farmer share of food prices has declined as well. The commission did not consider the development to be surprising, since most farm produce is

Table 4.1
National Budget Spent on Agriculture, 1950—1988, by Budget Categories within the Ministry of Agriculture and in Percent

	Percent of Agricultural Budget					
	1950	**1960**	**1970**	**1980**	**1985**	**1988**
Ministry	0.3	0.6	0.9	0.6	0.5	0.5
Rationa-lization[1]	9.8	11.9	13.1	5.3	5.3	6.8
Subsidies	67.5	51.7	34.2	67.7	56.2	55.8
Health, Quality Service/Control	3.4	5.8	9.6	3.2	2.8	2.6
Education Research	3.7	7.8	12.3	7.5	11.1	14.0
Environment	0.0	0.0	14.7	6.2	6.6	—
Sports	—	—	0	2.7	4.2	5.0
Other[2]	15.3	20.2	15.2	6.8	13.4	15.3
Total %	100	100	100	100	100	100
SKR 100000s	411.4	461.3	852.4	6,145.2	6,312.7	6,334.1
% National Budget	7.5	3.1	2.2	3.6	1.9	1.9

Source: RRV, 1950—51, 1960—61, 1970—71, 1979—80, 1987—88, Table 12.
Other includes expenses for forestry, fishery, and miscellaneous categories.
1. Includes contributions to the County Agricultural Societies, 1.4% of the entire agricultural budget in 1950 and 2.7% in 1960.
2. Includes contributions to the youth organization *Jordbrukarungdomensförbund*, .04% of the agricultural budget in 1950, and youth organizations in the forestry sector, .1% in 1950. The allocation financed organizational meetings.

processed in the post-farm sector. Yet the decline is a problem for many farmers whose economic position has tended to worsen over the years. Between 1950 and 1983, the farmer's share dropped 16 percent for beef, pork, cured meats, and eggs and 12 percent for grain other than wheat (SOU 1984:86, Table 5.19).

The food processing industry was another focus of concern. Various public spokesmen criticize the growth of the farmers' cooperative movement in the food processing sector. The commission viewed concentration of ownership and company mergers as a potential threat to Swedish war preparedness. It stressed the problem of Swedish vulnerability resulting from the internationalization of the food sector. The connection between the cooperatives and the food products under state regulation and the apparatus of agricultural corporatism has been the focus of special attention.

Several business endeavors of the cooperative movement have been reviewed by the NO, SPK, and MD.

Other topics taken up by the commission were consumer influence over all aspects of food policy and the role of agriculture for revitalization of rural areas and landscape preservation. The problem of surplus production was discussed in terms of its consequences for agriculture and the Swedish economy. That the public should help the farmers bear responsibility for surplus production was underscored in the report.

The report reflects the general political thinking of the time in other important ways as well. The unforeseen consequences of policy measures are discussed, and particular attention was given to the consequences for the environment from realization of the policy goal of rational and effective agricultural production.

A few of the commission members and experts submitted written reservations. The member who represented the Center Party stated in forceful terms that one conclusion from the report is that the commission aimed to phase out rather than develop agriculture. Two leaders of the LRF who participated as experts wrote a long, detailed, and critical statement in which they characterized the report as proposing a deterioration of the situation for farmers. It was obvious that a polarization of the interests of consumers and producers had taken place (SOU 1984:86, 491—518, 522—54).

Shortly afterward, the government presented its bill, which generally supported the conclusions reached by the Commission on Food Policy. The minister emphasized that the interests of consumers were the point of departure for food policy. He stressed that agricultural policy is only one part of a general food policy which focuses on the consumer. The quality of Swedish food, an issue of public concern and debate, was an issue of importance also (GB 1984—85, no. 166).

The interests of farmers in policymaking were not ignored completely. A modified exceptionalism characterizes the position that farmers were ascribed by the government. It had, when compared to its bill from 1967, changed its position on the policy goal of income maintenance and included it as an important aspect of food policy. Also, it admitted that the state should take a financial responsibility for the farm problem and proposed that it assume a 40 percent responsibility for surplus production of grain.

The bill marks a new and expanded social democratic policy stance. Considerations other than economic ones are singled out. The government and standing committee stressed that a development of agricultural policy toward greater consideration of the quality of food and the environment as well as increased regard for animal husbandry must take place in the near future (JoU 1984—85, no. 33). The general impression is that the Social

Democratic Party has accepted many of the criticisms of the traditional Swedish agricultural policy that were presented by the non-socialist parties in their bill in 1977.

It is not surprising that the decision in the Riksdag reflected a broader agreement. Only the Center Party voted against the general policy guidelines contained in the bill. Bloc voting occurred on a few policy issues, while on other issues the government received the Riksdag's unanimous support or support from one or two non-socialist parties (RP 1984 -85, no. 163).

Since 1985 general policymaking has been devoted to reform of agricultural policy. Another investigation was appointed in 1986 to study certain aspects of food policy that were highlighted in the 1985 Riksdag decision. They are the concentration of ownership in the food processing industry and the retail market, the geographic distribution of the industry, and the effects of the developments on food prices.

The committee was composed of only one member. A well-known consumer advocate conducted the investigation. Formerly he worked at the LO and represented it on the Consumers' Delegation. Later he became the head of the SPK and is now employed by the KF. Officials from the SPK conducted much of the investigation. An expert on agricultural and food policy and vocal critic of the market share of the farmers' movement at the SPK was appointed committee secretary. Only one person from the farmers' movement was included among the 19 experts associated with the investigation. It has been reported that the composition of the committee reflected an attempt by the government to formulate policy without the direct involvement of the representatives of sector interests, which were seen to include both MPs and interest organizations (interview with Ljusberg).

It is interesting that sharply formulated proposals and conclusions are not forthcoming in the final report (SOU 1987:44). Rather it seems that the report aims at sounding out the political actors. The government has decided not to present a bill based on the report. Rather, the conclusions drawn are allowed to influence further policy thinking.

As in other countries, it seems that politicians find it difficult to satisfy the different interests involved in food policy. The farm lobby demands that government assume responsibility for the unforeseen consequences of its agricultural policy. Consumers are concerned about the quality of food, high food prices, and other issues. Environmentalists demand change in agricultural production. Budgetary constraints and the general economic situation for agriculture limit the number of alternatives available. Politicians are aware that farmers and rural residents can articulate their interest through established interest organizations and that they can mobilize their vote. Policymakers have little incentive to advocate bold new approaches. Sweden is not the only country in the predicament (Olsen 1987, 1988; OECD 1987, 1988a, 1988b).

FARMERS AND ENVIRONMENT

The three traditional goals of Swedish agricultural policy—effective and efficient food production, war preparedness, and income maintenance—have been debated politically and their interpretation and respective importance has changed over the years. A good example is reformulation of the goal of food production. Initially, it emphasized the need for farms of a certain size and capacity for self-sufficiency. Now it aims at satisfying the consumer demand for high quality foods at low prices. As shown in this chapter, agricultural policy evolved gradually in the last three decades from a primary focus on farmers to an emphasis on the needs of consumers and society in general. Agricultural policy in the 1980s stresses another problem and reflects the second expansion of the policy arena.

Public concern for environmental quality and control increased in the late 1960s. Rachel Carlson's book of 1962, *Silent Spring,* dealt with chemical pesticides and is often cited as a catalyst for the Swedish environmental movement. Associations for the conservation and protection of the environment find their origins in the late 1800s and early 1900s.

The birth of the postwar environmental awareness can be dated to the early 1950s and identified to a handful of writers. An important forum for their thoughts was the magazine *Perspektiv*, published by the farmers' movement. The rural/urban cleavage was one of its central themes. One historian claims that it was a pioneer in the cultural debate (Frängsmyr 1977).

The environment became a salient issue in the 1960s. State committees investigated the issue, and public policy was formulated. By the 1970s, the issue ranked second after the issue of unemployment as the most important concern in the election campaigns. About 250 new laws on environment protection were enacted in the 1970s. The issue of nuclear power became salient in this decade also. A number of environmental groups were created at this time (Lundqvist 1971, 1980; Zetterberg 1978).

A public opinion for the greening of Sweden grew in strength in the 1970s. It has been claimed that the LRF newspaper *Land* introduced the concept of greening in the public debate. In Swedish it is called the green wave (*gröna vågen*). Numerous articles on the environment and chemical pesticides were included in the consumer section of the newspaper in the early 1970s.

In 1985, a national environmental party was formed. Environmental parties became active in municipal elections a few years earlier. By 1988 the Environmental Party received 5.5 percent of the vote and 20 seats in the Riksdag. It is also represented on over 90 percent of all local government councils and retains a pivotal position on 43 of them. By the later half of the 1970s, farm leaders in Sweden believed that the environmentalists ought to have less influence in politics and society, while the Center Party held the opposite view. Even the Swedish king has expressed publicly his concern for the environment (Verba et al. 1987, Figure 8.1B; Micheletti 1989).

One issue that gained priority in the public debate was the harm caused to the environment by modern agricultural production. Concerns for the environment became a fourth official goal of agricultural policy in the mid-1980s. New legislation on the issue was passed by the Riksdag in 1988. It is interesting that the minister of environment and energy rather than the ministry of agriculture, the traditional forum for reform of agricultural policy, initiated and formulated the government bill on which the legislation is based.

Table 4.2
Policy Issues Handled by the Riksdag's Standing Committee on Agriculture, 1945—1988, in Percent

Issue[1]	1945	1950	1955	1960	1965	1970	1975	1980	1985	1988
agriculture	75	82	83	92	82	69	67	51	60	48
tourism	1	1	1	1	4	4	7	8	8	2
environment	0	0	0	0	1	16	16	30	15	34
food quality	3	4	3	1	3	3	7	2	7	5
ruralism	11	1	0	2	0	0	0	0	1	2
youth	0	0	0	1	1	0	1	0	0	0
sports	0	0	0	0	0	1	0	0	0	0
other	10	12	13	4	10	7	3	7	8	8
Total %	100	100	100	100	100	100	100	100	100	100

Source: JoU, 1945, 1950, 1955, 1960, 1965, 1970, 1975—76, 1980—81, 1985—86, 1987—88.
1. *Agriculture* includes fishing, forestry, game, reindeer herding, farming, agricultural administration, regulation, and R&D. *Tourism* includes outdoor recreation, excepting sports, concern for the national parks, and the legal right to enter private land (*allemansrätten*). *Environment* includes issues of pollution, ecology, conservation, research, and administration. *Food quality* includes animal protection, taste, and general food issues. *Ruralism* includes issues of homestead policy, regional policy, and regional culture. *Youth* includes issues about young people that do not involve sports and tourism. *Sports* involves issues of organized sporting activities, sport clubs, and playing fields. *Other* includes miscellaneous issues, such as government lotteries, land surveying, non-farm land use policy, international cooperation, and procedural issues. The energy issue is also classified among other issues, as it is generally the responsibility of the Standing Committee on Industry and Commerce.

Environmental concerns have begun to dominate the agenda of the standing committee in the Riksdag. Table 4.2 shows that the issues of the environment, food quality, and tourism are crowding the agenda for the farmers. All are issues of postindustrialism. In particular, the environment issue is given priority. The chairman of the committee has estimated that 60 percent of its time is allocated to the environmental issue (Från Riksdag och Departement (R&D) 1988, no. 14).

What these changes point to is a greater number of interests and interest groups which compete with each other in the policymaking process. While the interests of farmers reigned supreme in the period of agricultural exceptionalism, today their organizations must confront other articulate groups whose interests reflect the new policy stream of agriculture.

ARE FARMERS STILL EXCEPTED?

Agricultural policy has been reevaluated over the past few years. Policy goals and guidelines are the subject of many public debates which focus less on traditional problems and allow for spokesmen to assess the problems in a new light. No longer is it solely the prerogative of the Swedish agricultural subgovernment. Other public bodies and actors seek to participate in the formulation of policy.

Criticism can be classified in two groups. There are critics who stress the costs of regulation and surplus production. The other group focuses on the issues of environment and food quality. Both concerns were stressed in the declaration from the new social democratic government after the 1988 election (R&D 1988, no. 31).

The ministry of finance has also become very active in the field as part of its plan to make the public sector more cost efficient. It decided to study the use of public finances for agriculture. Deregulation has been emphasized in different reports. A recent one proposed a new system for subsidies and other changes that would alter the situation for farmers in significant ways (Ds 1988:54).

The Agency for Administrative Development (*Statskontoret*) has also conducted investigations (1983). Its conclusions reflect the new doctrine of Swedish public administration, that is, the agricultural policy sector should manage its own financial affairs in a businesslike way. The National Price and Cartel Office and Office of the Parliamentary Auditors have voiced their opinions on agricultural regulation (SPK 1987, 1988; RRK 1979—80, no. 17).

The reform of agriculture is part of the general reform of the public sector, which has been formulated and implemented by the social democrats.

There seems to be a general agreement that state control and central economic planning cannot solve the problems of the 1990s: "The main buzzwords are privatization, deregulation, debureaucratization, reorienting of priorities and reorganization" (Olsen 1988, 1). The involvement of interest organizations in the process of government, bargaining model of governance, and negotiated economy have been questioned. The themes of the reform are political equality, economic efficiency, and individual integrity. Economic efficiency is to be strengthened by greater use of markets and competition, removing monopolies, and enhancing the influence of consumers and clients at the expense of producers. Overregulation by the state is stressed as a problem (Olsen 1988).

Another agency, the National Defense Research Institute (*Försvarets Forskningsanstalt*, FOA), presented a report in 1988 to the minister of agriculture that questioned the need for high agricultural production for war preparedness. It argued that vulnerability in the self-sufficiency in staple foods can be ensured through a network of international contacts as well as a higher level of storage of key products. Other issues of war preparedness were examined as well (FOA 1988).

In general, most public spokesmen criticize the complexity of agricultural policy and the role played by the farmers' movement in policymaking. Many believe that farmers ought to have less influence in politics and society (Verba et al. 1987, Figures 8.5B, 8.9B). They call for a simplification of the policy sector. Some focus their criticism on financial matters, others want a reorganization of the policy sector. All call for a policy reform.

When seen together with many of the changes in agricultural policy that are discussed above, it becomes obvious that the farmers must negotiate with a government that is sensitive to new problem areas and to the interests of the "nonfarmer." They are signs that point to the decline of agricultural exceptionalism.

Other more subtle changes are occurring that will affect thinking about agriculture. There is a widening of the polarization of the interests of consumers and those of producers. Fewer people are associated with the agricultural process, as a decreasing proportion of Swedes either farm or have rural roots. The majority of the offsprings of farmers have joined the working class, become consumers, and been exposed to a different political perspective on farmers. Employment opportunities perpetuate the gap caused by demographic changes. Also, there is less direct contact between farmers and consumers since food is transported longer distances (Wentzel et al. 1980).

Demographic changes are reflected in politics. Only 21 percent of all MPs in 1985 had fathers whose occupation was agriculture, and over two-thirds of the ministers in the 1987 government were from Stockholm (Holmberg and Esaiasson 1988, 137). Now only 17 of the 349 MPs have farming as their occupation (*Land* 1987, no. 3; *Land* 1988 no. 3). Tables 2.1 and 2.2 offer a historical account of the strength of agriculture in the Riksdag. A study from 1970 showed that over half of the middle and working class in Sweden did not have much contact with farmers (Steen 1980).

Table 4.3 illustrates how demographic change has affected the standing committee. Today only a minority of its members are farmers. More women participate in the committee. The Center Party is overpowered by the other parties whose political strongholds are not strictly rural or agricultural. Now the Greens are represented on the committee. An important issue addressed in Chapter 6 is how the farmers' movement attempts to counterbalance the changes in order to retain influence over policymaking.

It should not be concluded that agricultural exceptionalism had waned completely. The public debate on food quality emphasizes that food is not only a material product to satisfy nutritional needs but a cultural product that cannot be produced by technology alone. The role of the farmer and his view of the land is important here.

Table 4.3
Members of the Standing Committee on Agriculture in the Riksdag, 1920—1988, in Percent

Party[1]	1920	1930	1940	1950	1960	1970	1980	1988
Agrarian/Center[2]	6	19	13+	15*	15*	15*	20*	12*
Social Democrat	44*	38*	50*	50+	50+	55+	47+	47+
Liberal	25	19	13	20	20	15	13	12
Conservative	25+	25+	25	15	15	15	20	18
Communist	0	0	0	0	0	0	0	6
Environmental	—	—	—	—	—	—	—	6
Total	100	100	100	100	100	100	100	100
Farming as occupation[3]	69	82	75	80	85	80	33	41
Women	0	0	0	0	5	0	27	24

Source: Riksdag 1920, 1930, 1940, 1950, 1950, 1960, 1970, 1980, 1989.

1. Some of the parties have changed their names over the years. The names used designate their ideological position and approximate their current use. The Environmental Party entered the Riksdag after the 1988 parliamentary election.

2. The parties included in the category "agrarian/center" include both the group "*fria jordbrukare*," and the Agrarian and the Center Parties.

3. Includes hunting, fishing, and blue- and white-collar occupations in the agricultural sector.

* Includes the committee chairman.

\+ Includes the committee vice chairman.

Smaller-scale farming has reemerged as an issue. The implication is that it is difficult for anyone to favor agribusiness, a mode of production once supported by groups within the labor movement. Criticism of agribusiness is explained by the rural myth and the viewpoint that large-scale agriculture can neither ensure good food quality nor environmentalism.

The environment problem has in general cast new light on agriculture production. Farming that is more environmental is a salient political issue. Biodynamic farming receives public support now. Farmers who change their methods of production are now given a subsidy. The National Board of Agriculture has been put in charge of the issue. A council for biodynamic production has been created within its jurisdiction. Among its members are representatives for the interests of the consumers, the LRF, and organizations representing biodynamic modes of agricultural production. Agricultural schools as well as the University of Agricultural Sciences will begin to stress biodynamic production in their teaching and research.

A national association for biodynamic farmers (*Alternativodlarnas Riksförbund,* ARF) was founded in 1985, and there are a handful of local groups as well. The ARF organizes about 400 of the 700 biodynamic farmers in Sweden. A national institution which inspects and approves biodynamic production (*Kontrolföreningen för Alternativ Odling,* KRAV which means "demand" in Swedish) was created the same year. The KF, LRF, retail trade associations, and retail trade companies are board members. The KRAV is an institution for quality control that is neither public nor sponsored by the state.

Leaders of the ARF and KRAV state that they are disappointed with the response of the LRF, which they believe has the resources to support biodynamic production to a greater extent (interview with Carlsson, letter from Ahrén). The LRF president agrees with the assessment (interview with Dockered). Other organizations for biodynamic farming express similar sentiments and emphasize that the LRF has found it difficult to discuss new ideas. They see part of the problem being the wide range of interests among the LRF members, a situation which makes organizational development more difficult. The situation is said to have improved in recent years, as newer generations of farmers and elected officials become established (letter from Lundström).

Agriculture is neither neglected nor ignored in postindustrial Sweden. A new agenda for agriculture has developed, and policy expansion has occurred. Individual farmers and their organizations play important roles in implementing public policy on the environment, regional policy, and the concerns of consumers over food quality and prices.

The times have changed, and agriculture is changing to meet them. The interesting questions are how change occurs, what obstacles are involved, and which arguments are persuasive. Whether farmers, who were corporatized according to classical Swedish agricultural policy, can be convinced to partake a new task which requires them to support faithfully a new orientation in state policy is an issue adressed in the following three chapters.

5

Agricultural Corporatism

AGRICULTURE AS SAFEGUARDED GOVERNMENT INTEREST

Over the years, a great number of social and economic interests have been protected by the state. The number and kind of interests of the category differ among nations. In some as in Sweden, a larger number of interests are considered as legitimate concerns of state. This is one result of the welfare state. Once protected, interests receive a public status that involves the development of a bureaucracy which tends to their needs and defends them against other interests that compete for public attention and funds.

State involvement becomes both active and interventionist. Bureaucracy involves specialization, expertise, and organizational structure. Interests are protected, safeguarded, and developed in public policy. Sectorization of public policy results often. There are different consequences of sectorization, some of which have been the focus of criticism in the last two decades. They include policy rigidity, government overload, ungovernability, legalization, iron triangles, and the undue power of interest organizations in policymaking.

One of the first interests to achieve a protected or safeguarded status in government was agriculture. Chapter 1 considered why governments in most countries have identified agriculture with the interests of the state. The focus here is the Swedish situation, with the development and consequences of the agricultural subgovernment as its main subject.

The state began to be more active in the agricultural sector in the late eighteenth century. Yet recognition of the significance of agriculture for both the Swedish population and the state came over a century earlier. Creation of a central organization for the administration of agriculture was suggested in the 1600s, but it was not realized. An important change occurred when the basic structure of the government ministries was drawn up in 1840. Agriculture became the responsibility of one of the newly created ministries. The Academy of Agriculture was placed under its control and given more responsibility.

Debate in the Riksdag on a reform of the ministries in the 1850s put the issue of a special ministry for agriculture on the political agenda. The Estate of Peasants raised the issue. People outside the Estate began to put pressure on the government as well. Establishment of a ministry for agriculture was a pressing issue after the parliamentary reform of 1866.

A government bill argued that new ministries should be created for bureaucratic efficiency and effectivity. It was voted down. The parliamentary debate interested many MPs, and it has been concluded that no other proposal on reform of the ministries was debated in the Riksdag with such intensity (Thygesen 1925). Proposals for the establishment of a ministry of agriculture were raised in the Riksdag in the 1870s and 1880s.

The general mood or political stream—that is, public mood that crystalizes after public debate (Kingdon 1984, 152)—for a ministry changed a few year later. The county agricultural societies lobbied the government by submitting a petition on a special ministry to the king. It was argued that a joint ministry for different branches of industry would not be practical. The idea that agriculture could not be generally equated with trade and industry won sympathy. One of its first applications was the Riksdag decision of 1899 to establish the National Board of Agriculture in 1900. Agricultural exceptionalism was gaining support.

Not only did the event signify that agriculture would be treated differently from other branches of industry and business. It paved the way for the creation of an agricultural policy sector which meant that the state would take active responsibility for agriculture. The responsibility would come to involve different and related activities fitting into three general categories. Agriculture became part of the planned economy. The government decided that the national market of agricultural commodities should be protected in trade with other countries. Agriculture was to change to meet the demands of modern society with the help of the state.

Regulation characterizes the state efforts. Many regulations were economic—for example, public subsidies, taxes, and various methods for redistribution and improvement. Others involved physical planning (land use policy), the rules of policymaking (the regulation of regulation), and information (training, education, opinion formation, and R&D). All entailed the establishment of administrative units for their implementation.

Agriculture became a branch of industry whose economic health is almost completely dependent on public decisions and support. Periodic opinion surveys have shown that the Swedish public accepted the costs of state safeguards (Steen 1980). Peasant romanticism and its more modern variants, the greening of Sweden and rural rejuvenation, are important reasons for this.

Other explanations for the safeguarding of agriculture include the role that agriculture plays in Swedish war preparedness, sectorization of policymaking, and the nature of the government ministry itself.

Sufficient supply of agricultural commodities is an aspect of the security policy of most countries. As shown in earlier chapters, war preparedness has been one of, if not the central, fundaments of Swedish agricultural policy. Agricultural protectionism is a reason for Swedish food prices being much higher than international ones. Preparedness is implemented in peacetime through levies on imported commodities and through the acceptance of disequilibrium production, which in turn has a number of serious drawbacks for both the farmer and the national economy.

As long as self-sufficiency is accepted as an essential tenet of war preparedness, farmers are reassured that the public will support them economically and politically. The recent political discussion is signaling that a reevaluation is forthcoming. Attempts at reform in the past encouraged an outcry from several involved actors, including the organizations of farmers.

The significance of national security and national symbols as an underlying reason for the safeguarding of interests through state action and expenditure should not be underestimated (Stråth 1987, 20ff.). Perhaps this is the most important explanation for the lack of precise definition of the goals and means of the Swedish principle of agricultural war preparedness, for symbols are not conceived as belonging to the category of state activities which should be administered rationally.

Also, symbolism helps explain why all general reappraisals of war preparedness, which suggest that the level of Sweden's self-sufficiency in staple foods can be reduced, have been criticized relentlessly by the farmers and their organizations. Agricultural war preparedness positively identifies the Swedish farmer with national security and Swedish neutrality. Swedish farmers defend their country by tilling the land and, as in other countries, many are exempt from military service. National defense and protection of agriculture are seen as interrelated. The public discussion on Sweden's relationship to the EC and the GATT decisions on agricultural support systems signal a modification of the importance of war preparedness as a goal of agricultural policy.

For without a provision for war preparedness, border and market protectionism as well as the size, location, and commodity production of farms and the manpower needed for agriculture can be discussed from a different perspective or symbolic framing (Laumann and Knoke 1987, 314ff.), which expands the area of political debate on the issue, gives interest organizations either a competitive advantage or disadvantage, and leads inevitably to policy reform. It is argued in Chapter 4 that Sweden has entered this process for the second time.

Another reason for why agriculture is a safeguarded government interest is policy sectorization. The term refers to a situation where public policy is not made collectively by the entire legislative and executive branch. Rather, it is made by people whose expertise on the policy issue grants them a special role in policy formation.

Sectorized policymaking is often complicated and technical. Frequently, the participants—interest organizations with the heaviest stakes in the policy outcomes, legislative committees and subcommittees, ministries, public agencies engaged in policy implementation, scholars and journalists who are experts on the policy issues—consider the interests of the sector from the same point of departure and do not discuss the policy proposals in terms of general political development and the interest of the pubic. Neither do they attempt to make policy understandable to nonspecialists. Policy outsiders may find it difficult to apply a more general perspective since the policy is complicated and technical in nature. Their opinions may also be disqualified, for they are considered uninformed about the issues (Laumann and Knocke 1987, 129; Holmström 1987; Olsen 1987, 1988; Pedersen 1988).

A look at the government reports, bills, publications as well as scholarly books on Swedish agricultural policy will convince any member of the general public that agricultural policy is a quagmire of technicalities. In this sense, Swedish policymaking has been low track, though the debate on major policy proposals has been an agenda item for political parties, interest organizations, the mass media, and general public.

In many postindustrial countries, sectorized policymaking is associated with iron triangles, sponsored pluralism, and corporatism. The concepts are discussed in Chapter 1. Generally, they refer to "a fusion of the process of interest representation and policy implementation into a reciprocal relationship between the state and organized interests" (Cawson 1986, 18). Sponsored pluralism and corporatism are distinguished by the kind of interest organizations which engage in the political exchange with the state.

The Swedish situation is corporatist in that interest organizations that are countervailing powers are included in the exchange. The corporatist political exchange in agriculture safeguards the most concerned interests by ensuring that they are actors in policymaking. Unlike political exchanges of sponsored pluralism, the interests of the consumers play a part in agricultural policymaking in Sweden. Particularly social democratic politicians have striven to change agricultural policymaking to allow other interests to participate and become countervailing powers.

An important historical reason for the political exchange in agriculture is the active, interventionist state. The public administration does not simply administer rules. It implements policy, that is, causes effects and changes situations. It means that the state often must rely on interest organizations to help implement policy goals. Also, it signifies that interest organizations are delegated state responsibility and given the task of self-enforcement (*see* Offe 1985; Cawson 1986, 38).

INVOLVEMENT OF ORGANIZED INTERESTS

The state and interest organizations have aims or policy goals which they want to fulfill. The farmers' organizations want to represent the interests of farmers as best they can within the realm of state agricultural policy. Representatives of the consumers strive to counteract the influence of producers in the agricultural sector. The state seeks to implement the goals of agricultural policy that have been decided by the Riksdag. Each interest organization as well as the state has to decide how to maximize goal attainment under the circumstances given at any particular time. The interest organizations must also consider whether they can control others whose responses are needed for the attainment of their own goals.

The considerations are the prerequisites of rational social action (Dahl and Lindblom 1953, 54—91). Also, they form the two tiers of corporatism. One tier involves the political exchange or interdependence that was discussed in the above section. The other involves political accountability for the state, and responsiveness to member needs and opinions for the interest organizations (Katzenstein 1985, 33; Rokkan 1966).

Here the first tier is considered. The focus is the involvement of farmers' organizations and the representation of consumers in the agricultural subgovernment, and the consequences of their involvement. Chapter 6 discusses the second tier, that is, how the LRF represents its members.

The farmers's organizations are included in the policy process because their participation is seen as crucial for goal attainment. Representatives of the consumers participate because they are concerned interests. Wartime necessity and the Swedish principle of fairness and balance were the initial reasons for the participation of the interest organizations representing consumers. Later, their participation was seen as a way to change the emphasis of Swedish agricultural policy.

The major decisions on agricultural policy are all comprehensive reforms. They have changed organizational structures, processes, and performances. Unlike agricultural policymaking in some other countries, they were high track decisions: they had a high public visibility and interest, participants included a large number and variety of actors whose interest in the substantive content of the reform decisions tended to be polarized. Yet public interest in the reforms was probably lower than in other policy areas, labor and social policy for instance.

Frequently the goals of the major agricultural reforms were vaguely formulated and the means for implementation and control problematic. The legislation which was passed had generally the character of frame laws with the serious consequence that there may have been grave misunderstanding over the intent of policy, due to goal ambiguity (Olsen 1987; Daintith 1988). A close working arrangement or political exchange between the state and involved interest organizations was crucial in this situation if implemention were to take place. Low track decisionmaking characterizes the implementation stage in the policy process. In this respect, Sweden does not differ from other countries.

The political exchange between the state and the farmers' organizations in the agricultural policy sector occurs in different ways. First, both the RLF and SL and later the LRF have been important sources of expert knowledge for the state. For example, until the early 1970s the economic calculations for the price-setting negotiations were solely made by the JUI, which was the joint SL and RLF research institute. Statistics from the LRF are still used to set commodity prices, but they are now combined with other statistical sources. Experts from the LRF, along with those from the Consumers' Delegation, are members of permanent groups within the state unit, the Joint Council for Agricultural Economics (*Lantbruksekonomiska Samarbetsnämnden*), which compiles technical information on the economic situation of agriculture.

A second form of political exchange occurs when the organizations offer judgment on policy proposals by responding to remiss referrals. The government uses the reactions it receives in formulating its bill to the Riksdag. Another form is through participation in policymaking bodies, for example the state committees on agricultural reform that were discussed in previous chapters. The government may decide to sound out the farmers' organizations in less formal or public ways as well, that is, through consultations and negotiations. It conducted a series of secret talks on further agricultural reform with the political parties, the LRF, and certain other interest organizations including the representatives of the consumers a few years ago.

The participation of interest organizations in the formulation of policy is important for policy implementation. The acceptance of policy proposals by the farmers' organizations has been seen as an important step toward successful implementation. Acceptance by the organizational leaders signifies that they will defend the policy choices against criticism from the membership. Also they will inform the membership on the decisions and encourage their compliance, as cooperation by the members is necessary for successful implementation.

The reason for this is that the members are instrumental in the policy process since they more often than not are policy takers (Offe 1985). Political decisions are anchored within the interest organization in this way. Political exchange occurs in the implementing stage also. The farmers' organizations are frequently called upon to help implement policies. The interest organizations of the farmers are represented in all phases of the agricultural policymaking process.

The state views the cooperation and active involvement of the farmers' organizations in the implementation phase as crucial. In fact, it is probably the case that it sanctioned the participation of the organizations in policymaking in order to ensure that they will cooperate and help coordinate policy implementation. Examples of state reliance on the organizations from the war and inter-war periods are given in Chapter 3. Others are discussed here.

One example is the participation of the organizations in the rationalization of farms. Certain policy reforms require that the situation and behavior of the people who are affected by the policy (the policy takers) change. That this type of change requires a street-level bureaucracy with good rapport with the policy takers is substantiated by research from different policy domains (Rothstein 1986; Municio 1987). The main means for the implementation of the Swedish farm rationalization at the grassroots level is financial support and advice on farm practices (Nitsch 1972, 1977, 1982). It called for state reliance on the farm organizations, which took different forms. The expertise of the farmers' organizations was valued, and their membership on the governing board of the National Board of Agriculture is considered to be an important channel of communication and exchange of information (LBS 1986).

This type of contact has also met with the approval of the minister of agriculture (GB 1965, no. 100). The National Board has even encouraged the farmers' organizations to establish their own extension services. In fact it relies on their services since the county boards were not intended to and cannot meet the total service demand. Whether the state or the interest organizations should be responsible for this type of service was a question raised in the mid-1970s and is still a topic of debate (Nitsch 1977).

Probably the most interesting example of state reliance on farmers' organizations involves the county agricultural societies. Their development is both noteworthy and relevant for the present study for they cooperate closely with the LRF and will most likely play a more important role in agricultural policy in the future.

There are two main reasons for state interest in the county agricultural societies. The societies have a reputation of good contact with farmers since they have worked closely with farmers on an individual basis. There is a history of trust between the societies and the farmers, which is an important asset in any attempt at reform. Also, the county agricultural societies have always been closely supervised by the Swedish state. With the 1947 decision, the agricultural boards took over a number of the societies' most central tasks. They included the administration of state grants and loans. The county agricultural societies are quasi-public. Their constitution and tasks must be approved by the government, a task which is delegated to the National Board of Agriculture.

Until 1967 when the societies were reorganized again and more of their tasks transferred to the new agricultural boards, over half of their extension service was financed through public funds and about half of their employees were on the state payroll. The societies continue to perform a crucial function for the state. Through them the agricultural boards can distribute information to farmers. Their good reputation among farmers lends legitimacy to government information. The societies collect information on the results of policy implementation as well. In this sense, they function as a unit of feedback for the state.

Another specific example of state reliance on the farmers' organizations comes from the dairy industry. It involves the implementation of the two-price system for milk (*tvåprissystemet för mjölk*), which seeks to solve the problem of surplus production by paying dairy farmers a higher price for milk consumed domestically and a lower one (comparable to the international level) for surpluses.

The National Agricultural Marketing Board (JN) was made responsible for the implementation of the system which came into force in 1985. It requires detailed administration. The Board formulated the general guidelines and delegated the SMR the responsibility of operating the system. In fact, the SMR presented the Board with a proposal, which formed the basis for the general guidelines. It may be recalled from Chapter 3 that the initiative for milk regulation in the 1930s also came from the farmers' organization.

The economic associations which are members of the SMR sign production contracts with individual dairy farmers. Though not all farmers are satisfied with the production quota they are allocated, the majority of all dairy farmers (82 percent by 1987) signed contracts. Otherwise they must pay an extra fee on all their milk. In this sense, the system is not truly voluntary. The SMR set up a committee for administration of the system and appointed a board to grant exemptions and hear complaints. Criticism voiced by individual farmers who are members of the SMR are dealt with inside the organization. The state is only marginally involved in deciding controversies over production quota allotment; appeal to it has generally been impossible. This situation has been criticized by the NO (JN 1988a; NO 1989; *Pris och Konkurrens* (P&K) 1988, no. 3, 35f.; interview with Sahlström). As with dairy cooperatives in other countries, the SMR has played an important role for the dairy market. It has assumed the responsibility of relieving tensions among farmers caused by structural change (Baron 1978).

The system has been both a success and failure. Milk production has decreased. The number of milk cows as well as dairy farms and farmers has gone down. The system has led to desired structural change. Yet there are a number of unwanted effects: it is administered by a cumbersome bureaucracy, due process of the law is jeopardized since it is administered by the SMR and not a public body, and the economic development of dairy farms has been affected in a negative way. Recently, the Riksdag decided to abandon it. The SMR has been delegated the task of administering the deregulation, but the government has stated that the SMR may continue to implement the policy on its own without state help.

Another example of state reliance on the farmers' organizations involves attempts to solve the problem of surplus production of grain. A policy group (*Spannmålsgruppen*) was created in 1986 within the ministry to investigate the problem and suggest corrective measures. Members of the group include officials from the ministries of agriculture and finance, National Agricultural Marketing Board, National Board of Agriculture, Consumers' Delegation, and LRF. The LRF is the only interest organization with direct representation. Through its three representatives, the LRF has taken the initiative on a number of reform measures and helped formulate the general policy of the group, Adaptation 90 (*Omställning 90*).

It is a long-term program to decrease the production of grains and stimulate farmers to plant other crops. The goal of decreased production is attained by urging farmers to set aside a percentage of their grain acreage for other crops. The LRF has negotiated with the government on the program itself and the level of reimbursements to the farmers, who have voluntarily decided to sign a contract binding them to fallowing. It has also set its own goal—which is higher than the official one—for the number of acres to be set aside.

Though all its suggestions were not accepted, the LRF has campaigned for the reform within its membership and motivated its approval by referring to the seriousness of the problem. In 1988, 250,000 hectares were involved in Adaptation 90. The LRF has neither participated directly in the implementation nor been delegated responsibility. The Swedish Grain Trade Association, with the help of the county agricultural boards, is responsible for the reform program (DsJo 1986:2; DsJo 1987:2; interview with Herland). Support from the LRF has been essential. Information on the program is channeled through the newspapers belonging to the movement. Contact with individual farmers can occur via the local union clubs and county unions.

INDISTINCT BOUNDARIES BETWEEN PUBLIC AND PRIVATE

The corporatist exchange in the agricultural subgovernment of Sweden has blurred and weakened the distinction between private and public actors and responsibility. The cases discussed above are specific instances of this development. This section discusses briefly a few more general examples of how the public and the private spheres intermix in the agricultural sector.

The subgovernment is an issue network that involves interest organizations, companies owned by interest organizations, state agencies, companies owned jointly by the state, and interest organizations as well as quasi-public organizations. It is not very surprising that individuals involved in the sector can easily assume different roles.

Neither is it seen as worrisome that public and private legal authority overlaps in quasi-public institutions. A common legal form for them is the economic association. Economic associations are legally registered, and their activity is regulated in law. The state cannot create an economic association by itself, but it can own most of the membership shares and delegate administrative tasks to them (Marcusson 1986; Grosskoph 1981).

Examples of this kind of economic association are the 7 marketing associations for regulation of agricultural commodities. Though the state is not a member, it encouraged their creation and gave them the legal authority to perform the administrative tasks of market regulation. The state is represented on their boards, as are representatives of the consumers and producers. Public foundations are attached to the 7 regulatory associations. A public foundation is a form of collective ownership that is not easily distinguished from the economic association as both are nonprofit organizations.

Recently, the indistinct boundary between the state and the regulatory boards was investigated by a private management consultant firm commissioned by the ministry. A private firm was preferred over a public agency because many involved actors, including the National Agricultural Marketing Board, believed that a perspective from outside the sector was necessary (interview with L. Johansson). It is the same firm that helped with the reorganization of the Scandinavian Airlines System (SAS) a few years ago. The firm concluded that there are good reasons to separate the public tasks of the boards from their tasks of market regulation. It also reports that representatives for the state on the boards felt that they were in an inferior position both factually and psychologically when compared to the representatives from the farmers' cooperatives. Another conclusion of the investigation involved what it considered the unfortunate combination of democratic control and market mechanisms. It suggested a reorganization to remedy the situation (Indevo 1989).

A final example of the blurring of the public and private is a new public foundation recently created by the Riksdag for emergency disaster help to farmers. The government gave the LRF the responsibility of formulating the guidelines and financing and operating the help program, which also applies to farmers who are not members of the LRF. Responsibility for the older system for financial help after bad harvests will probably be turned over to the LRF in a few years. The LRF relies on people employed by the county agricultural societies to administer the program.

The creation of foundations that are held publically accountable is one means by which the government can deregulate agriculture without losing complete control over market forces or the involved interest organizations. Another method which will probably be tried in the future is the transfer of certain services that are performed by the county agricultural societies to the farmers' organizations (Statskontoret 1983; Johansson 1987). The services would become privatized but would still be under state supervision, due to public disclosure laws and the close contacts between the state and the organizations within the agricultural subgovernment.

In each cited example, the intermixture between the state and private interest organizations leads to the formulation of policy and the creation of institutions that are neither primarily accountable to the public nor to the members of the participating interest organizations. Another consequence is that it proves difficult for an actor to change its policy outlook or formulate a more articulated or adamant policy stance that takes its point of departure in the interests which the actor has been given the responsibility to safeguard. Change in policy must be negotiated among the policy actors.

BENEFITS OF EXTRA-MARKET FARMER POWER

The farmers' organizations have been given numerous opportunities to influence agricultural policy. Public policy has incorporated their interests. Also, it has furthered the exercise of indirect market power by the organizations. The ability of the organized farmers to exercise influence in this way gives them "extra-market power," that is, power that has not evolved solely from their position in the market (Bourdet 1988). Also, the concept of extra-market power helps explain why the productive organization of agriculture has developed differently from that of other business and industrial sectors (Koning 1983).

Swedish economists claim that agricultural regulation has reenforced the transformation of the agricultural processing industry into a cartel. It occurred because the state found that the monopolistic position of the farmers' cooperatives in certain key areas of commodity wholesaling was a satisfactory structural arrangement for the administration of agricultural regulation (Gulbrandsen and Lidbeck 1973; Rabinowitz, Haraldsson, and Bolin 1986). Leaders of the consumer movement generally agreed with the economists. They claimed that the state has favored the farmers' cooperative movement to the detriment of its competitors, for example, private companies and the consumer cooperatives in the meat processing sector (Bonow 1954).

The involvement of the farmers' organizations in agricultural regulation has benefited them in other ways. It has been found that the farmers' cooperatives have profited financially from the money received for performing public administrative tasks. For example, public money given to the LRF for certain tasks involved in hog breeding was also used for service to its members and the preparation of remiss replies on the issue (SPK 1984:2). The corporatist exchange on agricultural regulation has quite possibly given the farmers' cooperatives a competitive advantage in that they receive information on changes earlier than their competitors.

The farmers' cooperatives have attempted to use their extra-market power to influence market transactions. In all probability, other enterprises would choose the same strategy were they in the same situation. The most publicly debated example in recent years involved a cream substitute that a private company in the food sector (*Pågens*) introduced in 1981.

The SMR considered the product to be a threat to the whipped cream market. A problem facing the SMR in the 1980s was a surplus supply of milk, a situation caused by the gradual abolishment of the consumer subsidies on milk products. The subsidy on cream was discontinued in 1976. In 1982, it petitioned two state agencies in the agricultural subgovernment and asked them to take measures against the product. The product was taken up in the 1982 price-setting negotiation as well. The Farmers' Delegation of Negotiation argued that the protectionist fees on the product should be increased. A compromise was reached; the fees were increased but abolished later.

The reactions of the SMR to competition have been debated publicly. In particular, the Liberal Party criticized them. It has been claimed that the SMR wanted the product to be banned and that it used its extra-market power with this goal in mind (SPK 1984:20). The case led many opinion makers to question seriously the regulation of Swedish agriculture. The incident made an impact on policymakers and is cited frequently in reports from public agencies and committees.

REGULATION OF MARKET AND EXTRA-MARKET POWER

Over the years, the role that the farmers' organizations play in the agricultural subgovernment has been questioned in different ways. Representatives of the consumers were granted a more important role in the agricultural price-setting negotiations and on the commodity regulatory boards.

Much to the chagrin of the farmers' organizations, wholesale trading companies and business associations in the food sector, which do not belong to the farmers' movement, are now members of the regulatory boards. Since 1977 they have been granted representation on the Consumers' Delegation. Two seats are held by the associations. They represent the interests of consumers since the definition of who is a consumer has been broadened over the years in Sweden and other countries (Nadel and Vogel 1977).

The state can check the power of the organized farmers in other ways. Public agencies were created, and others were included in the subgovernment. The agencies participate on state committees and in remiss referrals on agricultural policy. They conduct their own policy investigations also. The Riksdag passed a number of laws intending to strengthen the legal position of consumers and protect their interests. Legislation regulates the operation and democratic arrangements of the cooperatives as well. A number of newer agencies safeguard the interests of consumers. They include the Consumer Ombudsman, National Board for Consumer Policies,

Market Court, National Industrial Board (*Statens Industriverk*, SIND), and National Food Administration. All were created in the 1970s.

Consumer influence is exercised in the governing boards (*lekmanna-styrelser*) of several agencies, as representatives from different interest organizations are included. The farmers' organizations and the interest organizations which represent the consumers and wholesalers are represented on the agencies within the jurisdiction of the ministry. It is noteworthy that the farmers' organizations do not retain seats on many of the boards of the consumer-oriented public agencies. They are not represented on the SPK, SIND, and National Board for Consumer Policies. Two of the critics of the present system of agricultural regulation, the LO and TCO, are. The trade associations are well-represented also. Repeatedly but unsuccessfully the LRF has applied for representation on the SPK board. Table 5.1 offers an overview of the interest organizations which are represented on the boards of the subgovernment agencies.

The growing strength of the nonfarmer interests in the subgovernment is evident in other ways. A survey of the agencies and interest organizations included in remiss referrals from the ministry of agriculture shows that the farmers' organizations have gradually become outnumbered in their own subgovernment. Over the years, the sphere of interest in agricultural policy has been broadened by the state.

The market and extra-market power of the farmers has been challenged and subjected to investigation by the public agencies that entered the subgovernment. A group of state agencies with the task of guaranteeing Swedish antitrust and industrial policy has been involved in this aspect of agricultural regulation.

Older agencies entered the subgovernment. The SPK, established in 1957, has increasingly become important in food policymaking. Generally, it emphasizes the interests of the consumers over those of the producers. In the 1970s, it conducted a number of investigations which criticized the market and extra-market power of the farmers' cooperatives. Its former director-general conducted the important investigation on food policy discussed in Chapter 4.

The NO has been given the responsibility of safeguarding competition in business. It began to monitor the farmers' cooperatives in the 1960s. The Competition Act of 1983 guides the NO in its work and gives it the special task of looking into limitations on competition that may be encouraged by state action, for example through regulation (SOU 1978:9; P&K 1982, no. 3, 5f.). It is the first agency to review whether company transactions limit competition in a dangerous way. If the NO finds competition to be in danger, the involved company or companies are requested to alter their

Table 5.1
Representation of Interest Organizations on the Governing Boards of Public Agencies in the Agricultural Subgovernment, 1988

	Interest Organization[1]												
	Producers			Consumers			Wholesale/Trade					Others[3]	No. on
Board[2]	LRF	SS	SLR	KF	LO	TCO	SI	GF	KMF	SLIM	Other[4]		Board
FS*											●	●	10
JN*	●			●	●			●				●	10
JTI*	●	●	●		●								13
KK					●	●					●		9
KV				●	●	●	●		●				12
LBS*	●				●							●	10
LBES*	●												8
NV					●							●	10
RSV	●				●		●				●		13
SCB					●	●						●	12
SIDA				●	●	●	●					●	14
SIND					●	●	●				●	●	11
SJFR*		●			●								13
SLU*	●				●							●	15
SLV*		●		●	●	●		●		●	●	●	14
SPK				●		●	●	●			●		10
SS					●								9
SVA*	●												6
VFN*	●											●	8
ÖCB					●			●					11
Total for organization	10	3	1	5	16	7	5	4	1	1	6	13	
Total board members													210

Source: Directly from the agencies.

* Under jurisdiction of the ministry of agriculture.

1. Except for the LRF, representation has been classified by peak organization. The LO and TCO member organizations are not included as their representation is defined as representation by the LO or TCO. The SI is the Federation of Swedish Industries (*Sveriges Industriförbund*); KMF is the Swedish Retail Federation (*Sveriges Köpmannaförbund*); GF is the Federation of Swedish Wholesalers and Importers (*Sveriges Grossistförbund*); and SLIM is the Federation of Swedish Food Industry (*Sveriges Livsmedelsindustriförbund*).

2. Union representatives on boards who represent the employees at the agency are not included. The FS is the National Board of Fisheries (*Fiskeristyrelsen*); JN is the National Agricultural Marketing Board; JTI is the Swedish Institute of Agricultural Engineering (*Jordbrukstekniska institutet*); KK is the National Board of Trade (*Kommerskollegium*); KV is the National Board for Consumer Policies (*Konsumentverket*); LBS is the National Board of Agriculture; LBES is the Joint Council for Agricultural Economics; NV is the National Environment Protection Board (*Statens naturvårdsverk*); RSV is the National Tax Board (*Riksskatteverket*); SCB is Statistics Sweden (*Statistiska centralbyrån*); SIDA is the Swedish International Development Authority (*Styrelsen för internationell utveckling*); SIND is the National Industrial Board; SJFR is the Swedish Council for Forestry and Agricultural Research; SS is the National Board of Foresty (*Skogsstyrelsen*); SPK is the National Price and Competition Office; SLV is the National Food Administration; SVA is the National Veterinary Institute (*Statens Veterinärmedicinska Anstalt*); VFN is the Board of Plant Hybridization (*Växtförädlingsnämnd*); OCB is the National Board of Civil Emergency Preparedness (*Överstyrelsen för Civilberedskap*).

3. Other interest organizations includes the national associations for fishermen and the Lapps.

4. Other trade associations include those which are specialized.

practices. If they do not, the Market Court becomes involved in the proceeding. The SPK aids the NO in its investigations.

Swedish cooperatives are not exempt from antitrust practices. Quite a number of transactions within the LRF's member associations have been reviewed. The NO has voiced criticism on three key practices within the farmers' cooperative movement. One is the principle of obligatory delivery of products to the cooperatives, which was included in the constitutions of many cooperatives. The principle of obligatory delivery has a long history. It was associated closely with the obligation of the cooperatives to purchase all crops produced by its members, an important principle that strengthened the market position of farmers before World War II. Earlier criticism of the principle by the NO in the 1960s led the SLR to change its constitution regarding delivery of grain and purchases of chemical fertilizers.

The NO's argument has been that a member of a farmers' cooperative should be able to choose whether s/he wants to deliver the commodity to the farmers' cooperative, a private company, or consumer cooperative. It maintains that reform of the delivery obligation will not harm the farmers' cooperative movement, for the movement has developed into the dominant form of enterprise in the food sector. The LRF's position has been that obligatory delivery is a necessary aspect of member solidarity since it does not want solely to become a selective incentive or service organization for its members.

The delivery issue was actualized again in the late 1970s. A law proposal for the Competition Act gave the NO a new basis to pressure the cooperatives to change their practices. The meat marketing cooperatives were focused upon. A compromise was reached between the NO and the LRF which meant that the delivery obligation would be still practiced, but members could decide to sell their commodities to another meat processer for a total period of one year. The NO agreed to accept the principle of delivery, and the LRF promised to recommend the change to its member associations.

The LRF cannot order its member associations to change their practices because they are independent units in the LRF federative structure. Not all the involved cooperatives followed the agreement made with the NO. One of the largest economic association within the SS refused. The NO petitioned the Market Court in 1985. The Court ruled against the farmers' cooperative, which subsequently changed its constitution (P&K 1986, no. 31, 46—56). Not many farmers have used the one year option. The NO plans to take up the issue of delivery again and desires more drastic reform this time.

A second area of concern for the NO in the early 1970s was the practice of the farmers' meat marketing cooperatives of retaining the money, that farmers deposited in them upon becoming members, once the farmer left them. Now the farmers are reimbursed one fifth of the sum over a 5 year period. Another area involved the practice of not allowing a farmer to sell beef, pork, or milk to a cooperative other than the one geographically assigned to him or her. The practice has now been changed (P&K 1986, no. 3, 46—56, P&K 1973, no. 5, 81ff.; NO 1981; interview with Sahlström).

The NO watches over the farmers' cooperatives in other ways. The farmers' cooperatives rate higher than any other business on the number of critical remarks from the NO (*Affärsvärlden* 1988, no. 39). The NO as well as the Market Court have been particularly concerned about the increased concentration of ownership in the meat processing branch to the farmers' cooperatives.

The NO plans to continue its monitoring of the farmers' cooperatives. It is an area given high priority. The NO believes that the principle of obligatory delivery that is included in the bylaws of the farmers' cooperatives should be restricted more severely, as the present agreement with the LRF is not judged to be satisfactory. Also, it has renewed its interest in the practice of many farmers' cooperatives to refuse repayment of the funds deposited for membership once a farmer decides to leave the cooperative (interview with Sahlström).

The general goal of the activity is to remove the internal regulations of the agricultural market by the farmers' cooperatives. The dairy cooperatives are singled out. Programs such as the two-price system for milk, which is a regulation of internal character, are seen as examples of how detrimental regulative politics in the agricultural sector are for the consumers and competition. They are "regulations that give birth to regulations" (NO 1988, my translation)

The SPK has received a renewed and strengthened mandate from the government to continue its work in this area. It is given a more offensive role in the monitoring of market competition. The SPK is commissioned to monitor and analyze competition in areas of particular significance to consumers and the national economy and where there is ineffective competition, which may now even include the public administration. International comparisons are encouraged as well. The name of the SPK has been changed to symbolize its more prominent role. It is now the National Price and Competition Board (*Statens pris- och konkurrensverk*), though the acronym remains the same.

GOVERNMENT REGULATION OF FARMERS

All aspects of the agricultural sector are regulated in one way or another by the government. There are food price, production, and quality controls. The export and import of most agricultural commodities are regulated transactions. There were several consumer subsidies; the one on milk is being phased out now. There are regulations for and subsidies on loans. Advice to farmers is regulated, as is land use. State agencies both set and review food prices. Farm incomes and budgets are regulated economies. The farmers' cooperatives are regulated by law. Within each of these general areas are quite a number of detailed rules and regulations (for a specification see Nilsson and Utterström 1988, 40—58).

A study conducted by the SPK shows that detailed regulation is characteristic of the agricultural sector. Together with the fishery sector, agriculture accounted for 211 statutes (*författningar*) in 1980 and 183 in 1986 (SPK 1987:3). Another count puts the ministry of agriculture at a middle ranking among the government ministries on the number of statutes and rules that were on the books in 1987 (GB 1987—88, no. 100).

The type of agricultural regulation that costs the most money for the state and taxpayers is subsidies, as shown in Table 4.1. The price-setting of agriculture commodities involves negotiation among the government, the National Agricultural Marketing Board, the LRF (Farmers' Delegation of Negotiation), and representatives for the consumers (Consumers' Delegation). In later years, a small farmers' organization (*Jordbrukarnas Riksförbund*) has petitioned both the LRF and the government for participation in the negotiations. Also, the LRF has requested a seat on the Consumers' Delegation. It reasons that the food processing industries that are cooperatives should receive a seat because the food processing industry that is privately owned is represented through its interest organizations. Both requests have been denied.

The negotiated economy of agriculture evolved over the last 4 decades. Different models of negotiation developed, and the general trend has been toward low track decisionmaking. The one that has been applied in later years involves negotiation between the Consumers' and Farmers' Delegations within the auspices of the National Agricultural Marketing Board. The Board sends a price-setting proposal to the ministry of agriculture. The ministry uses it to formulate a government bill, which is then discussed in the Standing Committee on Agriculture in the Riksdag. The Riksdag must vote on the proposal. It is uncommon for the committee to make important changes in the bill, and the Riksdag has never voted down a bill on the pricesetting of agricultural products. The prices of most food products are regulated in this way. The main exception is vegetables but not potatoes.

From the beginning, the Consumers' Delegation merely played an advisory role. Initially, its main tasks were to give the National Agricultural Marketing Board advice and information as well as to voice opinions on agricultural policy from the point of view of the consumers and retailers. A final task involved the distribution of information to consumers on Swedish agricultural policy. Here the general idea was that consumers should be given a better understanding of the policy and the farmers' situation: "to create a better understanding" (*att öka förståelse för*) or ideological control has become an important activity of all institutions in Sweden (Czarniawska-Joerges 1988).

Over the years, the LRF has requested that the government reevaluate the negotiating model. It desires to negotiate directly with the ministry, a model consciously avoided by the government. The National Agricultural Marketing Board has held the same general view as the LRF. In its comment on the proposal to create the Consumers' Delegation in 1962, the board defended the role that the farmers' organizations played in the price-setting negotiations and stated that the Consumers' Delegation should not be allowed to become a negotiating partner. It believed that there should be only two negotiating groups, the farmers and the state. Representatives of the consumers have tended to see such requests as substantiation that the LRF believes that farmers are in a special partnership with the state (interviews with Ljusberg and Nyberg). Also, the LRF has criticized the Consumers' Delegation for overly emphasizing the food price issue.

The LRF has stressed that the Farmers' and Consumers' Delegations do not have equal status. While their body is defined as a negotiating partner, the Consumers' Delegation is seen as an advisory unit to the National Agricultural Marketing Board. The distinction is emphasized in contacts with the government and in opinion formation campaigns. Most likely the LRF has, with due reason, feared that the organized consumers will become more influential in agricultural policymaking.

Since the early 1970s, the Consumers' Delegation has grown in stature. Though not codified in law, it has become a partner in the price-setting negotiations with the same status as the farmers' negotiating delegation. In 1978, its number of members was set at 11, an increase by one, and as stated earlier the definition of consumer representation was broadened to accommodate the heterogeneity of the interests of Swedish consumers. While the broader definition has met with criticism from the unions, trade and business associations think it should be even broader. Also, some of the members of the Consumers' Delegation have become professional experts and vocal critics of agricultural policy, a development which has been supported by the state.

The setting of agricultural prices has occured traditionally in two phases of negotiation. The first one proposes a total sum for the agricultural commodities under regulation and two food processing industries, the dairies and slaughterhouses, for approval by the government and Riksdag. The issues to be decided include how the total monetary amount should be collected: how much consumers should pay through higher food prices and how much the public budget should finance through tax revenues. The second negotiation concerns more detailed commodity price-setting.

The first and second negotiations are of principal importance. They can affect the direction of agricultural policy, the economic situations for individual farmers, and the relative burden for higher prices that will be shouldered by the consumers and taxpayers. The negotiations are an example of incremental policymaking and implementation. Issues that are related indirectly to the setting of prices can be negotiated. The participants in the negotiations can also be sounded out on policy reforms, an historic example of which is discussed in Chapter 2. Trial balloons are launched. The Farmers' and Consumers' Delegations are involved in both phases.

It is no wonder that the LRF takes the negotiations seriously. It uses its conflict fund to help form a positive public opinion for the farmers in the hope that it will have an impact on the negotiation rounds. Pep talks are held by the local union clubs. The central LRF leaders hold speeches and press conferences, and the administration produces material for the press. The research unit, the LUI, compiles and publishes statistics. Members of the LRF union equate a good settlement for farmers with the success of the leadership in representing their interests.

It has happened that the Farmers' Delegation has walked out and threatened to walk out of the negotiations. Also, it has demanded renegotiations. Acording to one newspaper, a threat by the LRF to terminate its cooperation in the secret talks on agricultural policy reform held before the 1988 parliamentary election led the minister to agree to a higher price for grain (*Svenska Dagbladet* 20 November 1987). Tractorcades have been held, and threats are made to conduct more of them. According to the present chairman of the Consumers' Delegation, who is also the TCO's representative on it, the threat frightened the minister into a compromise favorable to the farmers in 1988 (*TCO-tidningen* 1988, no. 10). Similar behavior on the part of Swedish public agencies before and during budget discussions with the ministries has been analyzed as symbolic performances, which are likened to rituals such as raindances and battles (Czarniawska-Joerges and Jacobsson 1989).

If negotiations become deadlocked, a mediated settlement is proposed by the Marketing Board. After the negotiations are completed, the minister submits a bill to the Riksdag which draws upon the settlement. The setting of commodity prices as well as policy measures are included. The minister may decided to raise policy issues in his bill. After discussion and possible modification in the standing committee, the bill is voted on by the Riksdag. Major changes in the settlement agreed upon by the two delegations have not been made, and policymaking here is more low track. Recent examples of important policy measures included in the price-setting bill include the decision to abolish the two-price system for milk and a lowering of the state's financial responsibility for surplus production of grain (GB 1987—88, no. 165).

While the price-setting negotiations have helped implement the four goals of agricultural policy through economic steering, other regulatory tasks involve administration. Important ones performed by the National Agricultural Marketing Board are market regulation and border protectionism, which aim at the maintenance of domestic food prices so that the price levels agreed upon in the negotiations can be upheld. The organization of war preparedness involves the Marketing Board, National Food Commission, National Board of Agriculture, and National Board of Civil Emergency Preparedness (*Överstyrelsen för Civil Beredskap*). Certain LRF associations are involved here as well.

The recent agreement on a gradual deregulation of agriculture and immediate stop to increases in agricultural prices by the GATT will change the institutions of price-setting in Sweden in a drastic way. It is likely that the agreement signals the termination of the price-setting negotiations between the two delegations as well as a reorganization of responsibilities involving market regulation and border protection.

A second important and broader concern for the subgovernment in the 1980s is the environment. Not only is environmental concern a policy goal with the same priority as the income and production (consumer) goals. It has become a significant area of interest for an increasing number of political actors in the subgovernment. People associated with the greening of Sweden are involved here also. Ecologists and people concerned about foods sprayed with pesticides, including producers and wholesalers, are two new actors in the subgovernment.

Probably the most important new state institution involved in the subgovernment is the ministry of environment and energy. Previous to its creation in 1985, the ministry of agriculture dealt with the environmental issue. The ministry of environment (as it is usually called) has been given the task of coordinating efforts on the environment within the government.

It is currently involved in formulating an environmental policy for the agricultural and food sector.

A maze of regulations is also developing for this newest agricultural policy goal. Economic steering through fees and financial supports is used as a regulatory tool, with the aim of either encouraging farmers to produce different commodities and farm within the new more environmental guidelines or discouraging them from farming according to the traditional postwar agricultural policy design. Fees on chemical fertilizers are used to finance agricultural environmentalism. Environmental considerations have been taken up in the price-setting negotiations, within the National Agricultural Marketing Board, and by the regulatory boards.

The National Board of Agriculture is another prominent agency in the agricultural subgovernment involved in the regulation process and the greening of agriculture. Its main task is service. In the past, it concentrated on advising farmers whose farm holdings fit the larger size norm of Swedish agricultural policy. Now the agency caters to all farm sizes. Its new directives include environmental concerns, rural development, and problems of surplus production. Biodynamic production is encouraged through its extension service and through financial support. Hobby farming on a small scale by urbanites is promoted, as the Board can apply land use policy to divide farms for allotment gardens. Considerations on recreation and outdoor life are two other important aspects of the work of the advisors now (LBS 1988).

Both the National Board of Agriculture and National Agricultural Marketing Board are involved in the quality control of agricultural products. The Marketing Board is developing economic steering measures to ensure a higher quality of food.

Concern for the environment is affecting the final set of institutions engaged in the regulatory system. They are the R&D units, including the Swedish University of Agricultural Sciences (*Sveriges Lantbruksuniversitet*) and Swedish Council for Forestry and Agricultural Research (*Skogs- och Jordbrukets Forskningsråd*). Both are under the jurisdiction of the ministry of agriculture, not the ministry of education.

The university has been criticized by a public agency for not focusing its attention on the environmental issue. The faculty of agriculture was criticized for its old fashioned view of the problems facing agriculture. The National Audit Board stressed that the state does not have the same interest in agriculture as it had in the past. It sees the role played by the interest organizations of agriculture in the university system and the isolation and sectorization of the university as a serious problem (RRV 1988).

SUBGOVERNMENT AND POLICY CHANGE

Since its conception the regulation of Swedish agriculture has been criticized in different ways. Critics in the 1980s have been the most vocal and influential. The debate on agricultural regulation and disequilibrium that is conducted in other countries is part of the reason for their persuasiveness. The constraints that are put on the national budget as well as the role of the countervailing power are two other important reasons. The explanations refer to the costs of regulation and the lessening power of organized farmers in the policy process.

Other reasons for reevaluation concern the changing interests of the Swedish consumer. Consumers are more urban today. Both family heads work, and this places specific demands on the food sector. Urban and rural life are developing isolated from each other, which makes it more difficult for the consumers to understand the situation of the farmers.

The policy, administrative, and legal considerations of regulation are other important reasons for reevaluation. It is questioned whether the four agricultural policy goals are compatible with each other and whether the means chosen lead to goal attainment. Detailed steering is characteristic of

Table 5.2
National Government Employees under the Jurisdiction of the Ministry of Agriculture, 1960—1988

	1960	1970	1980	1985	1988
National and County Boards of Agriculture	783	2,063	1,859	1,665	1,598
National Agricultural Marketing Board	*	148	151	147	137
Total in Ministry's jurisdiction[1]	5,097	5,760	8,072	14,448	12,539
Percent of all national government employees	2.2	1.9	1.4	2.4	3.0

Source: SCB 1969, Table 2; SCB 1971, Table 2; SCB 1978, Tables 1 and 2; SCB 1980, Table 2; SCB 1985a, Tables 1 and 2; SCB 1985b, Table 7.16; SCB 1988; and Kerstin Larsson, statistician responsible for the calculations at the SCB.

* Data not available.

1. The total number of employees under the ministry's jurisdiction has increased due to the expansion of its sphere of activity to include environmental considerations and conservation (until 1985), recreation, tourism, and miscellaneous tasks. Included in the statistics are also, for example, employees at the Swedish University of Agricultural Sciences and a number of specialized regulatory units.

agricultural regulation, as the jungle of regulations and their complexity has made it difficult to decide on more general guidelines for goal steering. Another set of reasons point to the benefits that the farmers' cooperative movement have received from the regulation.

Rarely is it pointed out that agricultural regulation does not involve a high investment of resources. It may be remembered from Table 4.1 that the proportion of the national budget allocated to agriculture is comparatively low, 1.9 percent in 1988. Table 5.2 offers other statistics on resources allocated to agriculture. The subgovernment does not have a very large staff in comparative terms. It is further evident that the two most important agencies for agricultural regulation, the National Board of Agriculture and National Agricultural Marketing Board, account for less than 15 percent of the ministry's employees. The statistics do not include employees of the county agricultural societies (roughly 1,000), the regulatory boards (about 80), and the subgovernment agencies not under the ministry's jurisdiction (see Table 5.1 for their specification).

6

Adjusting Farmers to Agricultural Change

HETEROGENEITY AND PROBLEMS OF REPRESENTATION

The LRF has two general groups of members. It represents individual farmers as a farmers' union and the cooperative business associations as a trade association. Discussion of the cooperative movement is reserved for a later section. Here the topic is how the LRF deals with the growing heterogeneity of its members.

Research has shown that it is more difficult to represent all members once the membership of an organization becomes more heterogeneous. Some of the differences between groups of members within the LRF are historical in origin. The different interests of large and small farmers as well as farmers in northern and southern Sweden are two sets of examples. It has been more common for large owners and southerners to farm on a full-time basis, while northerners and small owners tend to combine farming with other work, especially forestry. Other historical differences affect the representation of the farming interests as well. Party politics is one example. While small and medium farmers vote more often for the Center Party, large farmers and estate owners tend to vote conservative.

Also, significant differences resulted from the agricultural policy promoted in Sweden in the postwar period. Farm specialization increased as a result of the policy. The creation of clearly defined and formally organized subgroups or guilds (*intresseföreningar*) within the LRF is a consequence of the policy. The formation of guilds has led other groups of members to establish their own ones, a situation which the LRF leadership finds worrisome.

The most vocal, influential, and best organized guild is the one for grain growers. While the others do not have national breadth, roughly 3,000 large grain growers established a national guild association (*Sveriges Spannmålsodlare*) in 1987. It defines itself as a complementary unit to the LRF. It seeks publicity for its demands and has lobbied the ministry of agriculture. Nor does it avoid more militant tactics. It has opposed proposals for change in agricultural policy and criticized the LRF's efforts in the price-setting negotiations. The guild represents the traditional agricultural policy of Sweden's postwar period.

The LRF does not condemn the creation of guilds, though it is concerned about their potential militancy and the impression they make on the public. It believes that they jeopardize member solidarity as well. An example illuminates the problem. A transfer of money in 1986 from the Meat Marketing Board, which regulates the market, to help finance the high export costs of grain was criticized by livestock farmers. At roughly the same time, the grain growers' guild stated that it was unfair that grain farmers should be more responsible than livestock farmers for agricultural adaptation. Two years later, the grain farmers criticized a sizeable monetary transfer from the Grain Marketing Board that aimed at financing the relief system for dairy and livestock farmers.

A second consequence of heterogenity for the organization is the presence of members that criticize the LRF policies in an organized fashion. They are not guilds. Rather, they have more the character of discussion groups. A handful were present in the early 1980s. Each developed after a particular event, and all criticized the 1970 amalgamation (Bjärsdal 1986). They sought certain reforms within the LRF and wanted to convince local and regional leaders that more adamant action on agricultural policy was necessary.

The guilds and groups discussed here try to gain the attention of leaders on more narrow issues, which often have important implications for the LRF strategy and organization. They function as internal lobbyists, and their loyalty to the organization is not questioned seriously.

Other farmers have chosen to criticize the LRF more openly by joining a protest group. This is the third consequence of heterogeneity. The most notable is the Small Farmers in the West (*Småbrukare i väst*), created in 1984. Today it reports having about 1,700 members. Membership is concentrated to certain areas in western Sweden, though there are members throughout the country. The group states that it represents the interests of small farmers.

The Small Farmers in the West characterizes the LRF as a bureaucratic organization for large farmers, anti-environment and dominated by the cooperative businesses. It is not surprising that farmers who emphasize the value of direct democracy disagree with the LRF's actions and organization, which have developed together with agricultural corporatism. The criticism of representative democracy is present in other large Swedish interest organization as well (Micheletti 1985).

The group has received considerable and favorable attention, due in large part to the ability of its president to utilize the mass media. It says it has contacts with leading politicians from all parties except the Center Party, holds press conferences, and submits remiss replies. Yet it has not been granted legitimacy. Its request for a seat on the National Agricultural Marketing Board and on the Farmers' Delegation of Negotiation have been denied by the government and LRF respectively.

The LRF dislikes the group and the attention given to it. It has attempted to stiffle the group in different ways. The Executive Board directed that the organizational newspaper, *Land*, refuse advertisements from the group. The action was reported by a consumer-oriented news program, "Echo of the Consumer" (*Konsumentekot*), broadcast on the radio (18 February 1986). Public criticism of the LRF has benefited the Small Farmers in the West. Its populism appeals to those LRF members who view their organization as bureaucratic, undemocratic, and corporatist.

The cooperative associations have received their share of criticism as well. Farmers with a strong sense of individualism and instinct for business have decided to go their own way. They reflect two problems: consumers pay a high price for food while producers receive a low price for their commodities, and the cooperatives emphasize the production of quantity over quality.

Some livestock farmers have established butcheries on their farms as a way of solving the problem. The farmers' cooperatives are bypassed in this solution. One successful farmer markets "the healthy pig"; another breeds "the happy pig." They are slogans which reflect consumer concerns and the dominant spirit of the times. The food scandal of winter 1985 affected also the attitude of farmers toward the cooperatives. The radio program, "Echo of the Consumer," broke the story in one of its first broadcasts. The scandal was the lead story in many newspapers and news broadcasts; it involved the farmers' meat marketing cooperative, and dealt with the contents of livestock feed as well as sanitary conditions at the slaughterhouses.

It has been reported in *Land* that the price farmers receive for milk has prompted groups of farmers to submit their resignations to the SMR. In one case, an entire local economic association decided to leave the cooperative association because it found that the SMR was too large an association to consider its special needs and consumer demand for a particular cheese produced by it (*Land* 1980, no. 40).

Other lines of conflict are present, though distinctive groups have not evolved yet. Newly established farmers experience a tougher economic situation than farmers who have inherited their farms or who have farmed for many years. The situation is also quite different for full-timers versus part-timers. Other examples of potential lines of conflict involve the interests of female farmers and nonfarmer members.

In order to represent the interests of the heterogenous groups, the LRF must frequently either attempt to combine the different interests or concentrate on the interests of one group. An example of the latter has been the LRF's strategy in the price-setting negotiations and opinions on other issues, where it uses the interests of full-time farmers as the basis for its demands. Coping with the differences between groups of farmers and the organizational tensions which they create has become a major task for the LRF. It is crucial as the organization attempts to check declining membership rates by recruiting new members.

FEWER FARMERS BUT MORE OTHER MEMBERS

State agricultural policy affected the RLF membership in a very crucial way. Farm rationalizations led to a decrease in membership. State policy continues to have an impact on the membership of the LRF. The union and cooperative branches are both affected by it. Membership rates to the union dropped by about 8,000 between 1985 and 1988, as shown in Table 3.1. There is a decline in the number of farmers in Sweden which has led to a lower membership rate. But this is not a complete explanation.

In recent years, the LRF has lost a number of other members. Younger farmers and people who have become farmers recently tend to be more skeptical toward the LRF, which is reflected in their membership rates. The development may become very serious in the coming years, for roughly half of all farmers are over the age of 50 (Bondeson 1987).

Presently, the most serious trend is the lower membership rate among part-time farmers. Efforts to recruit them have not been successful. A survey conducted by the Agricultural Economics Research Institute (LUI) in the early 1980s showed that 83 percent of all full-timers (defined as working 30 or more hours on the farm per week) and 56 percent of all part-timers (working less than 30 hours per week) were members of the LRF (LUI 1987a). There are about twice as many part-time as full-time farmers in Sweden, though the latter group farms slightly over 60 percent of all arable land.

The LRF is concerned about the trend as the category of part-time farmers will increase in all probability. Another problem is that they as a

Table 6.1
Involvement of Farmers in LRF, 1987, in Percent

	Part-timers[1]	Full-timers
Membership[2]		
LRF	56	83
SMR	10	50
SS	43	78
SLR	53	82
SFF	53	73
SHS	10	42
SSR	36	39
Participation		
LRF activity[3] in 12-month period		
not once	78	52
at least once	9	12
at least 5 times	4	13
LRF activity once in two year period	17	51
Cooperative activity once in 2 year period		
SMR	6	29
SS	7	35
SLR	8	29
SFF	13	26
SHS	1	15
SSR	15	20

Source: LUI 1987a.
1. *Part-timers* are defined as farmers who work at most 30 hours per week on their farms. This includes both leisure-time farmers (less than 15 hours/week) and part-timers (15—30 hours/week). Leisure-time farmers tend to show lower involvement rates than part-timers. *Full-time farmers* are those who farm at least 30 hours per week.
2. See Table 3.2 for an explanation of the acronyms.
3. LRF activites include meetings, courses, etc.

group show signs of weaker solidarity. They participate less frequently in organizational affairs and tend to be less satisfied than full-timers with their ability to exercise influence, the information they receive from the organization, and the time spent at meetings. Also, the LRF is concerned about those part-timers who supplement their income by working in industry. Here the key issue is whether they will begin to show solidarity with the labor rather than farmers' movement. The involvement of part-timers in the organization is considered to be an important resource to strengthen the

LRF as an organization (LUI 1988). Table 6.1 summarizes the membership rates in the union and cooperative branches.

Recruitment of members is a high priority now. Too few members may mean that the representative monopoly of the LRF may be questioned. The LRF has decided to intensify recruitment among five groups: part-time farmers, women, the youth, farmers with forests, and rural residents. Membership for family members and sympathizers was approved by the RLF membership in the late 1960s as a way to increase its influence on rural issues. The LRF adopted the recruitment policy in 1975 but gave it priority only recently.

Recruitment of women was emphasized in the late 1970s. It has often been seen as an easy way to boost membership rates. By 1988 22 percent of the membership was female. Yet only slightly over 9 percent were elected officials, and very few were local union presidents (*LRF Aktuellt* 1986, no. 7—8, speech by LRF president in LRF Minutes 1988). Since 1977 there has been one woman on the LRF Executive Board. The concern shown by the LRF for women is explained also by the importance of women in rural society. As in other countries, the flight of women from the countryside is a serious problem.

Another reason for the LRF's interest in the recruitment of women is state policy on sexual equality. Every interest organization is now strongly encouraged by both the government and Riksdag to submit the name of one man and one woman to positions as its representatives on public bodies. The government chooses the most competent person. For the LRF it means that female members must be promoted and trained as competent leaders since no Swedish interest organizations wants to be labeled a male chauvinist in the public debate. Active participation of women members is seen by some as a benefit in its own right. It is hoped that women will be able to help revitalize the organization with new ideas and perspectives (LRF 1988a).

Efforts to entice the youth are motivated in the same and different ways. Rural rejuvenation and membership rates are important considerations, though a more specific problem is their declining interest in farming as an occupation. The economic situation facing many farmers discourages youngsters from agriculture.

Another problem is their low levels of participation as members. Attempts to solve the problem by the LRF youth organization (UK) have not been successful. A higher level of activity by younger members is crucial for the future of the organization, as older members tend to dominate the local and regional elected posts. Younger members are less likely to hold elected office and are, in comparison to their proportion of the membership, underrepresented in LRF decisionmaking bodies. To attempt to remedy the situation, the LRF will change its constitution to include a restriction against members over the age of 65 from holding elected posts (*LRF Aktuellt* 1984, no. 7—8; *Land* 1988, no. 38).

Finally, the LRF has much to gain from recruitment of forest owners and farmers who own forests. Both groups show lower rates of membership when compared to other categories of farmers. Only a small portion of all forest owners were members in the early 1980s, and the trend cannot be said to have improved. About 60 percent of all forest owners are not organized in the Swedish Forest Owners' Association (SSR) (*LRF Aktuellt* 1986, no. 6; *LRF Aktuellt* 1988, no. 4). Recruitment is important as forestry and forest policy plays an increasingly important role in Swedish agricultural policy. The problem is that the LRF and SSR do not have enough incentives to offer potential members, who do not need the help that the SSR can provide at present.

Successful recruitment of all the groups will increase the size of the LRF's membership. It will also satisfy members whose criticism of leadership recruitment policies has become more vocal. Yet broadening of the membership base in these ways will affect how the LRF represents its members, as heterogeneity will increase.

Critical voices have been heard already. Some members state that emphasis on recruiting rural residents will turn the LRF into a diffuse organization for the countryside. Others believe that it will be difficult to aggregate the demands of the heterogeneous groups (*Land* 1988, no. 8; LRF 1987a; *ATL* 1988, no. 8).

Certain groups feel threatened by the new recruitment policy. A survey showed that farmers with over 50 hectares and those who farm at least 40 hours per week were the ones least favoring the suggestions to organize and activate part-timers. Also, the group of full-time farmers tended to believe that the best way to strengthen the LRF is through more active participation of current members, who farm overwhelmingly full-time, as shown in Table 6.1.

No group viewed efforts to stimulate female participation as the best alternative, though a survey showed that over half of the membership felt that women should participate more. It is noteworthy that farmers under the age of 40 and part-timers found the proposal less attractive. The result is explained in part by the role that younger women play on smaller farms (LUI 1988).

CREATING SOLIDARITY

Implementation of the new recruitment policy is likely to aggravate the current problems experienced within the organization. Recent opinion surveys among the LRF members illustrate how difficult it is to satisfy all demands. One survey shows that there is a decline in the number of members who are satisfied with the LRF (LUI 1987b). The question is how the LRF can solve the problem, which more likely originates in the situation of farmers rather than with the LRF recruitment policy.

Attempts were made in the past to improve solidarity by encouraging members to become more active. All interest organizations have experienced that active members tend to be more loyal and satisfied with the organization than passive ones. Money is allocated to the local LRF unions to arrange events for the members, which vary from discussions on union and cooperative matters to social affairs. Participation of the members in organizational events is not very impressive, as shown in Table 6.1. The low levels give the LRF cause to worry. A knowledgeable membership is a prerequisite for the corporatist exchange and the formation of public opinion.

The goal of all member activity is discussion, debate, education, persuasion, and socialization. By talking complaints and criticisms over within the group, problems can be worked out and the members may be persuaded to see their complaints in a larger perspective. The perspective takes other interests of the organization into consideration and allows leaders to defend their actions.

To talk things over within the group is seen by all organized groups in Sweden as an important aspect of solidarity. Vocal criticism at group gatherings is acceptable. Criticism in an external setting is seen as disloyal. A central task for the interest organizations in Sweden is encouragement of an internal dialogue.

Every interest organization strives after a membership which, no matter how heterogeneous, has the same collective understanding of the importance of the institution and of the problems confronting it. This is an "organizational saga" (Clark 1972). This understanding cannot be assumed or taken for granted. Much of the spirit of the younger days of the organizations has disappeared. The situation is no different for the LRF, which uses the metaphor "to pull in the same direction" (*att dra åt samma håll*) to encourage members to learn and accept the LRF saga.

A measure that has become very important here is the information campaign. The LRF relies on new techniques from marketing, advertising, and pedagogy. The campaign is an example of ideological control, which targets the perception that members have of reality by offering attractive goals and convincing interpretations. Ideological control replaces the harsher methods of membership discipline that were used in the past. Most importantly, it gives members more of a choice (Czarniawska-Joerges 1988).

Various developments explain the LRF's reliance on ideological control.

The generally higher level of education in Swedish society renders the membership more articulate in its expression of demands and more capable to judge the leadership. Members must be convinced in a rational way that the organization performs its tasks successfully. Heterogenization of the membership can potentially isolate member groups from each other and inhibit solidarity. Ideological control tries to create an organizational saga that unites all groups.

Government and organizational decentralization forces central leaders to rely more on local and regional members who can argue for the organizational cause by adapting its policy to their own situations. The result is flexibility within the framework for the organization. Another development is the present political and economic environment for agriculture, which can cause members to lose faith in themselves and their organizations. Ideological control gives them a pep talk. Finally, members must increasingly be able to assess viewpoints that counter the organizational one. An organizational saga offers a point of departure for the assessment, and ideological control trains them to argue the LRF cause.

Ideological control is not only directed at the members. Discussion and dialogue is a crucial part. Campaigns offer the leaders an important opportunity to sound out their ideas, fly trial balloons, hear member suggestions and complaints as well as attempt to solve problems and formulate new ideas.

This type of activity on the part of interest organizations has become a booming business for consultants. Campaigns are frequently prepared by them and followed up by opinion surveys that measure member attitude. The role played by the LUI in conducting the surveys is important here. It is noteworthy that the LRF is the only Swedish interest organization with its own institution for opinion surveys; other organizations commission surveys from polling institutes and the bureau of statistics (SCB).

The campaigns that the LRF has given priority in the 1980s deal with the relationship between the cooperative businesses and the LRF union. The relationship is summarized in the metaphor, the LRF-model (*LRF-modellen*), which was coined in the early 1980s by the LRF chief executive officer. The message of the metaphor is that solidarity means belonging to both branches of the LRF and that the farmers' movement can only be strong when the branches stand united. The problems that the LRF-model attempts to solve originate in the history of the Swedish farmers' movement, that is, the different interests served by a farmers' union as opposed to a farmers' cooperative, and in the situation presently facing the cooperatives.

Problems did not disappear when the SL and RLF joined together to form the LRF. Some farmers express the sentiment that the cooperatives are bureaucratic and not fully democratic, a situation which the leadership does not deny. Others wonder whether the farmers' union has lost its spirit and potential force, as strikes against the cooperatives are not very feasible.

By the early 1980s, the leadership realized that it was no longer possible for the LRF to rely on the historic solidarity farmers once felt for the movement. Public sentiment in the 1970s for small, decentralized units added to member discontent, as supporters of the Center Party promoted the message that small is beautiful. At the same time, a wave of cooperative mergers occurred. Many members whose sympathy lay with the Center Party became alienated.

The metaphor "we and they in the organization" (*vi och dom i föreningen*) was coined in the early 1980s to summarize the problems accruing from larger organizational size, lower levels of member participation, loss of the spirit of solidarity, and the sense of distance between members and their leaders (LRF 1983).

The problem of solidarity between the union and cooperative branches has been debated since the early days of the LRF. The organizational newspapers carry articles, editorials, and letters on it. The general annual assembly and regional and local yearly meetings discuss motions on it. A few motions have proposed that the amalgamation be annulled. Investigation after investigation analyzed and tried to remedy the problem. Study circles are held, and the leadership has counciled with the members.

As shown in Table 6.1, there is a discrepancy in membership in the LRF union and in the cooperative associations. There are fewer members belonging to the farmers' cooperatives facing competition from private business. The government's plan for deregulation will encourage other businesses to enter the food sector. Though the situation may not appear as serious at present, there is a growing opinion among farmers that solidarity with the cooperatives is not very important.

The crux of the matter for individual farmers is how money paid for food should be allocated in the food sector. Some farmers refuse to deliver their commodities to the cooperatives for economic and ideological reasons. Others note that private wholesaling companies pay them a higher price. Here the meat marketing as well as the supply and purchasing cooperatives are affected in particular (SS 1986; Dahl 1987, 1988). Sometimes farmers see private companies selling them necessary equipment at lower prices.

Opinion surveys conducted by the LUI confirm the scattered reports of local criticism and actions against cooperative businesses. A survey from 1984 showed that the issues of democracy and equal treatment were concerns of the members. Members were satisfied with the service and contacts provided by the cooperatives (*LRF Aktuellt* 1985, no. 5). But when compared to an earlier survey, the situation has deteriorated generally (Utterström 1982). Also, they criticized their leaders for not doing more to solve the problem of surplus production. Younger farmers held the same attitude but tended to be more critical and were less apt to consider themselves as part-owner in the cooperatives. Negative attitudes among the youth are especially worrisome, as it is difficult to know whether they reflect inexperience with the movement that will pass or a serious trend.

It is evident that the farmers' movement is worried about the situation. Three LRF campaigns on these issues, which culminated in a member council in 1987—88 and a policy document, and two within the meat marketing cooperative have been conducted since 1985. Together the LRF campaigns go under the title the "Project on the Farmers' Cooperatives" (*Lantbrukskooperativa projektet*). The names given to the LRF campaigns quite nicely summarize their message and focus.

"A New Grip —Together" (*Nya tag —tillsammans*) was the name of a three-year project aiming to give the farmers' movement a new grip on life. Its topics included sounding out members on the LRF-model, member democracy within the cooperatives, the "we and they" syndrome, delivery of commodities to the cooperatives, and other issues of a more business character. Its goal was to improve the status of the farmers' cooperatives and the "we-spirit" (*vi-andan*), the metaphor used by many interest organizations to define member solidarity.

Study circles, meetings, and seminars are the forums for debate. Members and employees of the cooperatives participated in the circles. The adult educational association, *Studieförbundet Vuxenskolan,* was commissioned to help local units organize their efforts. Discussion materials and a film "Cooperation for Success" (*Samverkan för framgång*) were produced. They offer the member economic facts about the cooperatives, an organizational saga, and urge the members to consider the serious problems facing their movement (LRF 1985b, 1985c, 1986; board statement no. 36—39 in LRF Minutes 1987; Johansson 1988).

The LRF considers the campaign a success. It hopes that more members will understand the LRF-model and join both the farmers' union and the cooperatives. Approximately 10,000 people participated in the campaign, less than 10 percent of the membership. (There are roughly 20,000 people holding elected posts within the movement.)

Another step in the consciousness-raising for the LRF-model involved the appointment of a committee of high-ranking cooperative businessmen and elected officials to study the future of the farmers' cooperatives. The group was pointedly named the "Life Group" (*LIV-gruppen*), an acronym for its aim. Among its tasks was a critique of the LRF-model.

The committee found that there are severe organizational problems accruing from the 1971 joining together of the SL and RLF. Complexity increased as the roles and goals of the two branches became mixed. Tensions arose from the different interests of the union and cooperative businesses. Inflexibility characterized the LRF structures. The group discussed the increasing heterogenization of the LRF membership, and it took a stand for full-time farmers as the principle group of members for the cooperatives. The conclusion is noteworthy, given the current LRF policy to broaden its membership base. Their report became an important document for internal discussions (LRF 1987b).

Another activity was called the "Dynamic Dialogue" (*Dynamisk Dialog*) and aimed, as one might expect from its title, at encouraging people (over 60 experts and cooperative leaders) to debate organizational problems. A Swedish consultant firm was commissioned and used an American method developed to deal with difficult and systemic problems. The method is called policy delphi and is interesting in that it shows the constraints within the LRF organization. The method can be likened to outside help sought by a troubled married couple, whose problems involve difficulties of communication and disagreements on specific issues. The method attempts to develop constructive channels for talking and listening as well as an understanding of the problems. Terms from earlier decades capture its essence: a rap session, sensitivity training, and group therapy.

That an organization of primarily men, bureaucratic character, and traditional values chose to use the method says a lot about the perceived seriousness of the problem and the inadequacy of routine methods—including the study circles, debates, and meetings with members and leaders—to deal with them. That policy delfi emphasizes the anonymity of written responses in its first stages reflects the general inhibition toward vocal criticism felt by both leaders and members. Communication within the organization is a top-level problem. A report commissioned from another consultant firm stressed this same point (Relationskonsult 1987).

Later the group of leaders who participated in the dynamic dialogue met to discuss the viewpoints articulated in the questionnaires. The project took over one year. The effort is an example of how a large interest organization attempts to regain some of its youthful vigor over half a century after its initial creation.

Many of the topics taken up in the dialogue focused on the future of the cooperative businesses. Concern was voiced over the multiplicity of interests as well as the delivery of commodities. It was stressed that today's members must be treated as individuals and that they should perceive membership in the cooperatives as personally advantageous. The LRF president stresses that farmers should choose membership in the cooperatives for economic rather than solidaristic reasons (interview with Dockered; LRF 1988b; *Nya Perspektiv* 1987, no. 4, 9f.).

An important metaphor coined in the early 1980s attempts to deal with the problem of heterogeneity, growing individuality and solidarity within the cooperative movement. It is "usefulness to members" (*medlemsnytta*), which summarizes different reasons for membership in the cooperatives. It combines the collective and selective incentives for membership and emphasizes that membership cannot be based on material, short-term, or immaterial considerations entirely.

Examples of positive collective values are benefits from the corporatist exchange, service to all members, and the importance of a strong cooperative movement in harsher times. A significant example cited frequently by the LRF is the obligation of the cooperatives to buy members' commodities when the market is saturated, when it is in a situation of surplus production. It is hoped that the metaphor will convince members that membership will outweigh short-term economic losses and that they should deliver to the cooperatives, though a private company may offer them a better price (LRF 1986, speech by CEO in LRF Minutes 1987).

Persuasion of the membership is important since the provision of selective incentives cannot be used as a main argument for membership. This is because the LRF allows farmers who are not members (nonmembers) to purchase many services at the same price. Though they are nonselective services, the LRF believes that services are an incentive for membership. They include insurance policies and tax consultation. The LRF section for consultation includes seven units, a staff of 2,000, and 170 offices throughout Sweden. It is the largest consulting agency in Sweden. The LRF argues also that the services provide a necessary source of income for the organization. A few more minor services are true selective benefits, and special economic offers for new members are being developed as well (Motion no. 50, board statement no. 50 in LRF Minutes 1988; *Land* 1988, nos. 38, 49).

Perhaps the predicament experienced by U.S. farmers' organizations has made an impression on the LRF. The political influence of a number of organizations diminished, as policymakers suspected that farmers were not committed members and only joined the organization to obtain the selective benefits of membership (Browne 1982). Traditionally, interest organizations in Sweden have not emphasized selective incentives as an argument for membership. The situation is beginning to change now.

The metaphors the "LRF-model" and "usefulness to members" played a central role in the member council "Agriculture Makes A Choice" (*Lantbruket väljer väg*) conducted in 1987—88. The council was the culmination of the campaigns mentioned earlier. As implied by the title, the issue of membership was one of its explicit themes. An implicit theme is the role of leadership in business management and elected office.

It is interesting that many large interest organizations in Sweden are focusing now on improving the capability of their leaders. The reason is the importance of contacts between members and leaders to counteract the feeling of anonymity within the membership. Here the LRF does not differ from other large interest organizations. The leaders are given the responsibility of promoting solidarity while satisfying the interests of individual members at the same time. Also, they are the bearers of the organizational saga.

A study conducted by the Meat Marketing Association offers an excellent exemplification of the problem. It stressed that Swedish agricultural policy and the harsh economic situation for livestock farmers led to a concentration on bureaucratic expertise and a deemphasis of member issues, which had a serious consequences for the local leadership. As their role became less significant, they felt a sense of inadequacy that affected their leadership ability in a negative way. It contributed to the declining member solidarity: delivery of livestock to the cooperatives declined from 85 percent in 1969 to 77 percent in 1986 (SS 1986).

Enthusiastic and capable leaders will inevitably be very important in this regard once agriculture is deregulated. Leadership recruitment is an important concern. Here it is crucial that the leadership reflect both the viewpoints of the membership and their composition. If the lines of conflict within the LRF are not to develop into serious problems, leadership must be representative, democratic, flexible, and effective. Rotation in office is important as well. Steps have been taken to encourage members to seek leadership posts. No longer do farmers suffer economically for the time devoted to the tasks involved with elected office. But active participation by members who are not leaders is not renumerated, a situation explaining perhaps some of the nonactivity among certain groups of members. Attempts are made to remedy the situation. One target is young and new farmers (SS 1986).

The most important reason for the search for leadership is the ongoing decentralization within the LRF organization. A clearer relation between responsibilities and influence is sought as well. Flexibility in the movement is to occur through decentralization, and it is hoped that this will remove the stigmatization of the farmers' movement as bureaucratic and undemocratic. Also, the reform should save money for the LRF. The Meat Marketing Association summarizes its approach with the terms: principles, flexibility, and closeness.

It is too early to evaluate whether the ideas will lead to growth in membership for the union and cooperative movements and a better solidarity among the members. Earlier campaigns conducted in a more traditional fashion have not improved the situation. Campaigns conducted by the Meat Marketing Association were criticized for their propagandist content and said to be merely "wordy rubbish about solidarity," as the "usefulness to the members" metaphor was never specified (SS 1986).

FARMERS' COOPERATIVES IN MARKET CAPITALISM

Cooperatives in Sweden are undergoing fundamental change. As is the case in many other countries, they have reached a third phase of development. In the first phase, they were created to help members with problems. For example, the effects of the buyers' market led to the creation of Swedish farmers' cooperatives in the early 1900s. The second phase is service to members, as exemplified in the intense efforts of the Swedish associations to provide protection for the members.

Now cooperatives are becoming market-oriented. It is a result of advances in capitalism (LRF 1986a). The farmers' cooperatives have been forced to adapt to modern market capitalism and consider consumer demands in order to retain their position in the food processing sector. Modern capitalism signifies that firms compete over market shares, clients, and even values. This creates environmental uncertainty for the cooperatives (Daudi and Sotta 1986).

One serious consequence is that it has become more difficult for the associations to assert and adhere to basic, traditional cooperative principles. The necessity of adaptation has led to a debate among scholars and within the cooperative organizations. Words like "ideological crisis," "lack of identity," and "schizophrenia" summarize the viewpoints of certain scholars of the cooperative movement (Røkholt 1986, 1987). Others stress the negative effects of monopolization on the farmers' cooperative movement (Bolin 1987). A third group views the cooperative as an ideal type, which implies that deviations from the ideal are expected (Nilsson 1986). All scholars focus on the issue of cooperative renewal. Here the two important aspects are willingness to adapt to market capitalism and the role to be played by the cooperative principles.

Cooperative businesses are frequently seen as having more complex roles and goals than private firms. They are interest organizations which safeguard the interests of their members. They are actors in a competitive market. Frequently they are delegated government responsibility and play the role of a public administrator. Dairy cooperatives are the best example here; the Swedish experience is discussed in Chapters 3 and 5.

The different roles complicate the strategy of action for cooperative businesses. As market competitors, the businesses must emphasize market mechanisms, efficiency, and effectivity. As interest organizations, they must give thought to the effects of market strategy on their members. As public administrators, they must take the public interest into account when deciding market strategy and representing the interests of their members. Cooperatives in Sweden have a social responsibility, as they are accountable to other groups in society, and they are considered as an important social movement as well. The Swedish farmers' cooperatives play an important role in agricultural change, rural development, and environmental policy. They play a central role in the Swedish economy and its food sector in particular.

Together the cooperatives belonging to the LRF are the largest owners in the food sector. They form Sweden's third largest enterprise and are involved in both primary and secondary food processing, though they are less dominant in the latter. They employ around 50,000 people (a decline by 18.5 percent since 1979). Many cooperative business are in rural areas. As mentioned previously, an economic association within the SMR, Arla, is the largest firm in the Nordic countries. They can be called "supercooperatives" (Young 1987), as indicated in Table 6.2.

It is not surprising that there is a general concern for the cooperatives in Sweden. The government appointed a state committee in 1977 to study the role of cooperatives (SOU 1981:60). A government bill on capital accumulation in the cooperatives was presented in 1983. Concern over their future led also to a number of campaigns within the LRF. Their central issue is how a balance should be struck between cooperative solidarity and principles and market forces.

The issue is whether the farmers' cooperatives are losing their character as cooperatives. The problem is aggravated by a tendency for the enterprises to adapt their practices to those of other businesses, lower solidarity for the cooperatives among farmers, and a decline in the number of farmers.

Surplus production and the costs of processing foods have led the cooperative enterprises to rationalize their operations. Three waves of mergers occurred earlier; a fourth is underway now. Many farmer members and the unions that represent the employees at the cooperatives are critical of the mergers. For farmers, the issues are the level of service and the influence of members in larger associations. For employees, it is their jobs. Politicians are concerned as well. Their issues are food quality and regional policy.

Alternatives are considered to improve the financial situation of the cooperatives. The issue of capital accumulation has been stressed in the

Table 6.2
Market Shares of the Swedish Farmers' Cooperatives, in Percent

Fertilizers	80
Feed	86
Farmer insurance	85
Farmer loans	86
Machinery	40
Milk	100
Slaughter houses	78
Grains	80

Source: Adapted from Bolin, Meyerson, and Ståhl 1986, Table 2.2.c

1980s. The movement learned from its bad experience in the 1970s—when many members belonging to the forestry cooperatives lost financially—that solvency is a central aspect of economic soundness. Presently, all cooperative associations are quite solvent, with 25 percent as the minimal goal. The introduction of cooperative enterprises in secondary food processing on the stock market is a second alternative that has been tried. The cooperative banks became a stock company in 1986. The LRF general annual assembly approved this strategy in 1988.

Many companies owned entirely by the farmers' movement are stock companies. The LRF owns 14; all are engaged in providing services to farmers and the cooperative associations. When the LRF was constituted in 1971, it owned only 7. Other companies providing service are owned by the LRF together with its cooperative asssociations. Also, the LRF and its member associations have invested in stock companies involved in research and development, as the development of food items and new industrial products from agricultural commodities is seen as important for the future of Swedish farmers.

Together the cooperative member associations own entirely over 50 stock companies. They are primarily service, R&D, and secondary food processing companies (*Land* 1988, no. 19). The creation of stock companies out of dairies within the milk marketing cooperatives has attracted attention, as it deviates from the general LRF principle that the cooperative form of enterprise should be the type of enterprise that lies closest to the members.

It is felt that market capitalism and cooperative principles are not completely compatible and that attempts to adapt them to each other create problems. One example is a SMR-owned stock company (*Semper ab*) that is a member of the Federation of Swedish Food Industry (*Sveriges Livsmedelsindustriförbund,* SLIM). The SLIM is a trade association that is rather adamant in its criticism of agricultural regulation. It alternates with other associations as a member of the Consumers' Delegation. It agrees with critics who point out that the monopolistic strength of the milk marketing cooperatives and their inflexibility to market signals is promoted by agricultural regulation. It criticizes the LRF directly.

As the cooperative enterprises merge to strengthen their economic foundations, another issue of importance has been their role as employer. Here it is clear that developments within private business have affected the cooperative enterprises. A cooperative "profit-sharing" is seen as a method that can increase the attractiveness of the businesses on the highly competitive labor market and encourage employees to feel a sense of solidarity with farmers. It has even been proposed that employees be allowed to become members of the cooperatives, a suggestion that the current membership finds dissatisfactory. Yet the ideas are included among the strategic goals for the farmers' cooperatives that were approved by the 1988 assembly.

Competition and cost consciousness is encouraged among farmers as well. The viewpoint supported by the LRF leadership is that farming is a vocation, not a lifestyle. The LRF service units help farmers to create their own specialized cooperative and stock companies. Examples include groups of farmers who own machinery together and who have jointly transformed their farms into stock companies. Companies are owned by small groups of farmers whose purpose is to purchase equipment and supplies at a lower price than offered by the farmers' supply market cooperatives.

As such ventures promote cooperation and solidarity among farmers, they tend to be supported by the LRF. Also, they minimize the financial risks taken by individuals, an important point considering the economic hardship experienced by many farmers. Perhaps business cooperation among farmers may provide a safety net for farmers whose debts make them undesirable members for the established cooperatives. In one controversial case, the supply and crop marketing cooperative in northern Sweden canceled its deliveries to farmers with bad credit ratings (*Land* 1980, no. 18).

Yet it is hoped that the metaphor "usefulness for members" will be developed into a simple economic calculation that can be used to compare cooperatives with each other and private firms. The *Land* attempted to operationalize the metaphor in a series of articles on the economic soundness of the cooperatives in the same branch (*Land* 1988, nos. 11, 16). Competition between cooperatives involves a comparison of their professional status and the prices and services they offer to individual farmers. In certain areas, it is possible for the farmer to choose which cooperative s/he will sell the products.

Competition within one cooperative concerns higher prices paid to certain farmers for particular commodities as an incentive for membership. The meat marketing cooperatives that implemented the strategy have been supported by the LRF. Other examples that are more controversial include higher prices for higher quality and lower charges for large purchases.

The crucial issue here is reevaluation of the basic cooperative principle of equal treatment of members. The subject is debated within the farmers' movement today. The recommendation from the LRF to its member associations is that the principle cannot be applied successfully without consideration given to the competitive nature of the association's market (LRF 1986b). The general conclusion is that principles should neither be bound by history nor tradition. Rather, they should be evaluated in the context of the market and can differ among member associations. The question of part-ownership in the cooperatives rather than membership for farmers is relevant here, though serious discussion has not really begun within the movement. Yet if adopted the cooperatives will, to a larger degree, be likened to private stock companies.

A central aspect of the new strategy involves more coordination at the central level and less associative independence. Federalism within the Swedish consumer and farmers' cooperative movements means that disagreements between the units responsible for the entire movement and those whose responsibility is limited to a particular branch or region can occur. Some scholars view federalism and its potential conflicts as a prerequisite for organizational democracy. But many of the policy issues of the past that were specific to a particular cooperative branch are now important for the movement as a whole (LRF 1985b, 1988b, 1988c; Lundholm, Hagbergh, and Utterström 1982; Svensson 1985).

More authority to the Farmers' Economic Company has been investigated and discussed within the LRF for over a decade. A report accepted by the LRF general annual assembly in 1979 on the LRF's role in this process was never implemented. The Plan of Action for the Farmers' Cooperatives (*Lantbrukskooperativt handlingsprogram*), accepted at the same assembly, has not had a significant impact. Both stressed in carefully worded formulations the need for coordination in a changing market environment (LRF 1979a, 1979b).

Greater coordination will most likely be necessary as Sweden grows closer to the European Community. Presently membership in the EC is not advocated by most political actors. The LRF has debated the issue, and it took more time than expected to formulate a standpoint. The LRF is not forthright in its opinion on membership in the EC. Rather, it has decided to wait and see how the debate develops while preparing itself for greater trade in agricultural products (LRF 1989).

Emphasis is also placed on the possibilities available for Swedish products in the European market. The catch phrase is "Good Food from Sweden's Farmers" (*Bra mat från Sveriges bönder*), which is written in Sweden's national colors. The concept includes marketing of environmental (biodynamically grown) products, an activity supported by the LRF only recently. It may be the case that the farmers' movement plans to follow the export strategy of other Swedish industries, whose aim is to find a small niche in a large competitive market by producing a specific product of reputable good quality.

No wonder that the LRF decided to emphasize the marketing over the production of agricultural products. Consumers are slowly replacing farmers as the target group of the cooperative enterprises. The recent campaigns within the farmers' cooperatives focus the attention of individual farmers on the importance of economic considerations in planning for the future. They are to become aware of the role that consumers must be allowed to play in food policy and adapt themselves to it.

Three of the strategic goals that were formulated after an evaluation of the campaigns discussed earlier and approved by the LRF general annual assembly in 1988 are increased market-orientation, internationalization, and development of new food items. What they have in common is the hope that the modern reasons for the existence of a Swedish class of farmers will be accepted by society at large. One of the strong reasons for the acceptance is satisfaction of consumer demand.

SHARED INTERESTS OF FARMERS AND CONSUMERS

The general hope within the agricultural policy community is that broad agreement can be reached on food quality, rural and environmental concerns as well as surplus production and deregulation. They are the key issues in the current discussions of agricultural reform and reflect the different constellations of interest involved in the policy domain. This section takes up the consumer-orientation of the farmers.

There have been two main consumer demands on Swedish farmers: food at a low cost and high quality. The important issue is the rising proportion of family income spent on food, a situation more severe in Sweden than in other western industrialized nations.

Concern over the quality of food has been expressed in various ways over the years. First it meant nutrition and reasonable prices: every citizen should be able to eat a balanced diet. Later quality became synonymous with low prices, quick preparation, and health (sanitary conditions and low fat). The second phase coincides with the rationalization and effectivity thinking of the 1960s and the early greening of Sweden. Farm production, food processing, and retailing was large scale, standardized, and hygenic. Quality defined as taste entered the discussion at a later date. The present view of food quality is a combination of the considerations and three additional ones, animal care, the environment, and choice of food products.

Concern for food prices and quality within the LRF reflects the development. Current emphasis on the desires of consumers is explained in three ways. Consumers have become a countervailing power in agricultural policymaking. The reorientation of the farmers' cooperatives to market mechanisms means that the preferences of the consumers will be more in the center of focus. This takes into consideration the push toward deregulation, trade expansion, and import penetration, as reflected in the discussions on membership in the EC and the developments within the EFTA and GATT (Bourdet 1988). Finally, the farmers' movement appeals via corporate advertisement to consumers in order to create goodwill, as a way to remedy their faltering image.

Swedish society has become more heterogeneous in its composition and attitudes. One implication is that consumer preferences in food and food quality have become less standardized. There are consumers whose main demands are satisfied with the production policies of the 1960s and 1970s. There is a significant consumer group—the "postindustrials"—that places other demands and is willing to pay for better food quality, defined as taste, variety, and assurances that food is produced environmentally. The group is also the generation that has traveled extensively and developed a preference for other types of food.

The farmers' cooperatives are acutely aware that many consumers are not loyal buyers. They purchase what they like to eat or can afford, without concern for whether they buy imports. Their national consumer loyalty is low. The cooperatives know that food scandals affect the trust of all consumers toward the food processing industry. Purchasing power is becoming a means to influence the food market.

The farmers' cooperatives are developing a competitive consciousness. In an important sense, they are now being forced to show more loyalty to consumers, a situation that differs from the one experienced in many other countries (Grant and Coleman 1987). Purchasing farm production is gradually being replaced by taking account of consumer preferences and transforming them into marketable products as the primary goal of the cooperative food processing industry. For the first time, cooperative leaders are becoming attuned to classical market signals.

Campaign after campaign within the LRF membership stresses the central role that the market and consumers must play in the future. Their aim is to convince the members that the strategy is the best one. The current position of the LRF leadership is that consumer preferences must govern the actions of the farmers' movement. Consumer concern over pesticides and animal care are changing the operation of the cooperative enterprises (LRF 1985a, 1985b, 1985c, 1986c).

The LRF general annual assembly accepted a policy document on quality control in 1985. Its title is "*Prisa kvalitet*" which in Swedish can mean praise, award, and pay the price for quality. Emphasis is put on concern for the environment, product differentiation, consumer-orientation, and national consumer loyalty (LRF 1987c).

Farmers are encouraged to produce higher quality commodities, and higher prices are paid for quality production. Another goal is the formulation and enforcement of branch policies for better quality. For example, the meat marketing cooperatives have adopted a policy for animal care that gives more consideration to physical and psychological animal health. An award is presented yearly to a farmer who practices and a researcher who develops quality care.

A main goal is protection of the interests of farmers and the cooperatives through creation of national consumer loyalty. Consumer loyalty to domestic production is seen as requiring public relations. Though economists tend to disagree on the importance of corporate advertisement, they agree that businesses invest in it to help create consumer loyalty to impede other firms from entering a particular market (Bourdet 1988). Advertisements paid by the cooperatives offer consumers information on products, quality control, changes, and alternative ways to use the products.

The Swedish Dairies' Association in particular has devoted many resources to advertisement. Billboards in the subway stations and around town as well as commercials in moviehouses show well-known people—politicians, singers, athletes, actors, and actresses—drinking milk products. The idea is to convince the youth that milk is as prestigious as the soft drinks advertised by rock stars.

Other efforts attempt to convince the public that drinking milk is important for one's health. Celebrities jump up and down to illustrate the importance of milk as a source of calcium. Recently the minister of environment took the jump, as this critic of the environmental practices of farmers is reported to have promised she would when the association stopped using bleached paper in its packages. The jump is an important symbol for the farmers' cooperatives. Another campaign is entitled "Milk is Life" (*Mjölk är livet*). Activities such as sports events are sponsored by the milk marketing cooperatives. The advertising media is used to focus the attention of different groups of consumers on Swedish food and farmers in other ways as well.

FORMING PUBLIC OPINION TO COPE WITH POLITICAL CHANGE

Two basic problems complicate the task of public opinion formation for the LRF. Its nature requires an organizational strategy that encompasses the interests of farmers and the cooperatives. Lower numbers of farmers in Sweden mean that there are fewer farmers and rural residents as policymakers in all sectors of society. No longer can the farmers' movement rely on automatic support from MPs and leaders in government. The LRF president summarizes the development by stating that the hegemony of farmers in Swedish society has been replaced (speech by Dockered 1989).

It has taken the LRF members and leaders a long time to realize that their political capital is lower than in the past. An understanding of the interests of farmers is a subject that must be learned, as fewer people are associated with agriculture now. One consequence is that the LRF must lecture the public on matters of importance for agriculture. It has become a teacher of sorts. Over the years it has developed an impressive organization to accomplish the task.

Information on the plight of farmers and their movement is seen as a crucial step in the exercise of influence. Attempts to influence take the form of public opinion formation and the corporatist exchange. They aim to improve or maintain the status of organized farmers within the food issue network.

They can do this because politics does not unfrequently involve conflict over the definition of the problem and political stakes. Different groups in society have their own interpretations of which facts are relevant for policymaking. The interests at stake in a conflict are often an important issue as well. A good example is the disagreement over who (producers, consumers, or rural residents) agricultural policy is to benefit primarily.

The implication is that material benefits are not the only focus of political parties and interest organizations. All political actors are aware that they must win "characterological struggles." It involves agitation and conflicts directed toward the establishment of a group's interests as legitimate social and political demands. This is necessary to receive the tangible outputs of policy (Gusfield 1963).

Group interests become politically legitimate in different ways. The two examples emphasized in this book are farmers, whose status derives from their role as food producers, and consumers, whose position as a countervailing power was proclaimed by the state. There is a constant struggle between groups for social and political position, for their status is never stable. When society changes, new arguments to maintain group status must be formulated. The farmers' movement formulates new arguments now for the characterological strugle over agricultural policy. War preparedness is less influential as an argument for agricultural exceptionalism. Food quality has gained in importance as a good argument. The LRF has characterized its weak position in the public debate with the phrase "pitfall of opinion" (*opinionsfälla*). The general problem facing the farmers' movement involves concrete policy or position issues as well as expressive or style issues (Merelman 1969). An example of the former is the commodity price-setting negotiation results. The future of the rural myth is an example of the latter.

Opinion formation is a complicated task. Problems arise when both style and position issues must be encountered and the interests of the union and cooperative branches satisfied by the same basic message. Analysis of different LRF opinion formation campaigns shows that the organization uses and blends market and peasant romantic arguments. Given the heterogeneity of its membership, the choice is not surprising.

Purer market arguments are frequently present in connection with the price-setting negotiations. They aim to convince consumers that the farmers cannot be blamed for high food prices and to put pressure on parliamentarians to give the interests of farmers more consideration. An alliance with consumers is sought. Information is distributed via various channels. Articles appear in the consumer section of the *Land*, which circulates to over 400,000—about 11 percent of all households in Sweden or slightly more than the largest morning daily. The success of this section of the *Land* is an excellent example of the public relations ability of the LRF.

Special brochures are also produced for distribution to consumers. They are published in tens of thousands of copies. One with renewed publication is entitled "Where Does All Money For Food Go" (*Vart tar matpengarna vägen*). Price statistics, which often discredit the results presented by other groups, are calculated for journalists and other agenda-setters. The LRF also advertises its economic message in daily newspapers. One campaign in 1985 portrayed a full-page picture of the prime minister kicking a soccer ball or the minister of finance pushing a grocery cart. A catchy title and economic arguments on food prices were presented in bold letters on the adjoining full-page page.

Books discussing food price and quality issues are published by the LRF publishing house (*LTs förlag*). Short films, commercials, and local radio programs are a high priority. Press conferences and information as well as more traditional political campaign PR materials (bumper stickers, buttons, memorabilia, and billboards) are produced. A more traditional forum for debate of current issues is the "Agricultural Week" (*Lantbruksveckan*) which has convened every year since the early 1900s.

Words, more often than physical acts like strikes, are the union tactic for negotiation. Over the years the LRF members have gladly appropriated additional money from their union dues for campaign purposes. There are over 90 million Swedish crowns in the conflict fund that is primarily used for the characterological struggle. (There about about 6.5 Swedish crowns per U.S. dollar.)

Evaluations conducted by the LUI and other consultant firms tend to show that the money is well-spent. An evaluation after the consumer campaigns in 1981 and 1982 show that roughly half the population noticed and a fourth read the advertisements (*LRF Information Aktuellt* 1982, no. 6). Members have motioned on measures to improve contacts between the mass media and the LRF. Though courses on agricultural issues for journalists began in 1985, many members believe that journalists need more education on the issues. They tend to perceive that the media has become more negative toward them, a viewpoint seconded by many LRF member associations.

The importance of institutional advertising is reflected in the trend for member associations and cooperative enterprises to establish their own information and press units rather than rely solely on the LRF's capabilities. The 5 largest associations have their own units now. Courses for union members on how to communicate with journalists have been developed also.

The farm problem is given a lot of coverage in the mass media. The plight of individual farmers is focused upon. A surprising number of radio and television programs have taken up the life stories of individual farmers who often farm small plots of land. They are not the member groups that the LRF tends to emphasize.

A form of information combining romanticism with realism attempts to lessen the psychological and physical distance between consumers and farmers. It personifies farmers and encourages the general public to associate agricultural change and criticism with particular individuals. Also, it attempts to convince the public that farmers are not complainers dependent on government handouts but serious businessmen and good family members. Examples include art exhibits with farmers as the motif, art scholarships, books on farm and rural life published by the *LTs förlag*, and events to encourage consumers to meet with farmers. The LRF leadership has formulated metaphors to accentuate this effort.

Contact between producers and consumers is judged to be crucial. Some of the events are conducted in the cities, for example at Skansen and at fairs. A museum that is a farm with both modern and primitive farm methods exists and had 10,000 visitors in 1985. A slaughterhouse museum opened recently in Stockholm. An important opinion formation event is the "Open Farm" (*öppen gård*) and consumer day activities that started in the 1980s.

Such activites were available to more than half a million consumers in 1987. Consumers are invited to visit a farm and discuss issues with the farm family. The role of individual farmers in the characterological struggle should not be underestimated, and many members have attended courses where they learn debate techniques.

The evocation of empathy for farmers in the general public is the aim of the activities. The LRF realizes that politics is neither completely rational and realistic nor irrational and romantic. Rather it combines these reactions (Elder and Cobb 1983). Peasant romanticism became an important part of the official LRF message in 1985 (interview with Ramberg).

Farmers have also become highly visible to urbanites through demonstrations. As discussed in Chapter 1, tractorcades are common in several countries. The Swedish demonstrations are also necessary for organizational solidarity. They unite different groups of members by focusing their attention on the state and giving them the opportunity to vent their frustrations as union members. Demonstrations are crucial in this regard because the LRF's exercise of influence in the corporatist exchange is low track and outside the sphere of member contact.

The national demonstrations that occurred in 1971 in Jönköping, the location of the agricultural agencies, and 1985 in Stockholm, the Swedish capital, each gathered over 20,000 farmers and received the attention of the mass media. They dealt with the price-setting negotiations. A film company owned by the *LTs förlag* produced a video film on the 1985 demonstration, which has been likened to the Farmers' March of 1914. More local actions in protest of price settlements demonstrations have taken place as well. The TCO and particularly the LO were the brunt of a protest in 1987 because they presented a proposal for lower grain prices.

Civil disobedience and unconventional methods for the formation of public opinion were not disapproved explicitly by the LRF leadership a few years ago. A truckload of grain was sprayed with a chemical pesticide and was dumped outside the door of the ministry of agriculture in February 1986. A tractorcade was timed to block a motorway from Stockholm to ski resorts during a school holiday. Farmers have also considered burning surplus grain production, an action reminiscent of French farmers' dumping of milk.

Very few actions are directed against primary food processors since they are farmers' cooperatives. Raising member consciousness on the LRF-model helps to steer protest actions away from the cooperative businesses. One of the few exceptions involved the SLR, which was prohibited in 1988 from unloading a ship filled with imported rye. It has been reported to the author that the action was implicitly approved by the LRF leadership and that the SLR received compensation from the LRF for the boycott, though it was not known whether compensation came from the conflict fund under the jurisdiction of the LRF union. The LRF hopes that the rye boycott is viewed as a poignant example of current contradictions in state policy, which prohibits Swedish farmers from using certain chemicals but allows the import of commodities sprayed with them.

Recently a number of protest actions have occurred that have not been sanctioned by the central LRF leadership. Angry farmers have dumped their milk, blockaded a dairy, the National Board of Agriculture, and National Agricultural Marketing Board, and confronted the minister of agriculture when he appeared in public. The actions are a poignant criticism of agricultural reform. Though directed at the state, they criticize also the participation of key LRF leaders in the adaptation toward a postindustrial agricultural policy.

CRUSADE FOR THE HEARTLAND

Research by political scientists has shown that interest organizations, though representing a fraction of the population, can sucessfully further the interests of their members in the political process. The few can defeat the many by restricting or broadening the scope of relevant issues. A useful way to manipulate the scope of conflict is through symbols. A general law of power was formulated by a political scientist in the 1940s: "When men want power, they act according to their expectation of how to maximize power. Hence symbols (words and images) affect power as they affect expectations of power" (Laswell 1965).

Though style issues are important for all social movements and interest organizations, they are most crucial for those institutions that feel threatened and consider themselves underdogs. Dramatic mechanisms that use symbols to communicate to the general public are useful for organizations that are heterogeneous, for they play upon the organizational saga and help reach a common understanding among the members. Political issues which divide former allies will also encourage their use. Dramaturgical techniques are necessary for organizations that attempt to renew their public legitimacy (Merelman 1966).

No wonder the LRF has increasingly chosen to rely on the symbols of nature to reenforce its status. The choice is wise when consideration is given to the tenuous bonds between rural and urban areas, the degeneration of the countryside, the stronghold of peasant romanticism, and the current impact of postmaterialistic values. The crux of the issue is how the values of traditionalism, modern business methods, and postmaterialism can be articulated in a coherent program.

The set of values associated with nature is difficult to formulate in economic terms. It confuses policymakers and interest organizations that were created when Sweden was an industrial rather than a postindustrial nation. The problem is complicated as well as risky for institutions that develop the countryside; values traditionally inherent in developing the countryside cost money. It seems as though institutions are sounding out each other and the general public on their dedication to the countryside.

Opinion surveys have shown that the public attitude is more positive toward farming than in the past. One reason given is the role farming plays in preservation of the cultural landscape (Drake 1987). The mass media plays an important role in the process as well. The key issue is whether the national mood will swing to support an agricultural policy with more emphasis on nature, that is, a postindustrial or -materialistic policy. The LRF is watchful and supports its less controversial policy proposals and measures.

The campaign, "All Of Sweden Shall Live!" (*Hela Sveriga ska leva!*), is the Swedish contribution to the campaign "Let's Make the Most of the Countryside" initiated by the Council of Europe a few years ago. The LRF was involved in the campaign organization composed of about 90 Swedish popular movements and interest groups. The rural myth is the ideological base for the campaign. The organization that the LRF is associated with uses the slogan "Where Angels Tread" (*Ängla Mark*), the title of a popular Swedish song. The campaign aims at sensitizing the general public to rural and farm problems and formulating policy measures.

The LRF has been successful in using the campaign for its own purposes. It considers the campaign an important strategy in two ways. The propaganda potential of the campaign is tremendous. The campaign was used as a weapon against the government, TCO, and LO. Involvement in the campaign is good for member solidarity. It allows them to discuss a common interest and participate in activities. The LRF leadership used the campaign successfully to instill a brief sense of pride in its members. Farming is time and again proclaimed as the heart of the economy and community of the countryside (speech by CEO in LRF Minutes 1988; statement by LRF president in *Land* 1987, no. 24).

Concern shown by the farmers' movement for the cultural landscape is neither new nor solely propagandist. Along with other involved interests in the late 1960s, the RLF was a member of a committee that dealt with protection of nature in times of farm rationalization. For many years, the LRF has concentrated on rural culture. A program was accepted by the assembly in 1974. One of its aspects was the cultural landscape. A more specific program on the issue was commissioned by the 1982 assembly and accepted in

1985. A new perspective on rural culture was presented in 1988. It emphasizes the importance of a cultural hearth for the farmers' movement anchored in folk history and oriented toward the future. The goal is to offer a viable alternative to urban culture. The identity of farmers is at stake (LRF 1985a; cultural policy in LRF Minutes 1988).

Other efforts to crusade for the heartland are clumsy, unsophisticated, or overly cautious. A recent national opinion formation campaign conducted by the LRF concentrated on advertisements in the three largest cities. The means of communication were full-page ads in newspapers and billboards. It combined verbal proclamations with national romantic art. Ads included in the campaign related the following vague messages: Farmers are soft and automobiles are hard; the farmer is among the 15 species of life in Sweden threatened by extinction.

The LRF defended the campaign from criticism by stating that the ads reflect the self-irony and humor of the farmers' movement, and they provoke the general public. It is a good example of corporate advertising by a Swedish popular movement. An opinion survey showed that the ads were read by a considerable portion (49 percent) of the urbanites interviewed and that they felt relatively positive toward the method of advertising (40 percent) (Arne K. Larsson 1988). "Any attention is better than none at all" can be said to be the motto of the movement. It is doubtful that the ads will win a professional marketing award, as occurred in 1986 for a campaign entitled "A Countryside That Is Alive Needs An Agricultural Sector That Lives" (*En levande landsbygd kräver ett levande lantbruk*).

A second set of values dominates in the debate on the countryside. They are economic. A policy for rural areas is formulated in a regional policy of economic orientation. Employment, depopulation, and infrastructure are the main topics. The traditional actors in the agricultural policy issue network feel comfortable with the problem complex, where value is defined as productive capital. The concerns were found to be important for rural residents as well (Glesbygdsdelegationen 1987).

The LRF has been quite active in formulating regional policy by participating in state committees and formulating its own standpoints. In 1986, it circulated a regional policy program commissioned by the assembly three years earlier. The report discusses in cautious language the conflict of interests that can arise within the farmers' movement over support to sparsely populated areas. Cooperative mergers are seen as necessary for financial reasons. Yet they can make it difficult for farmers whose products must travel long distances to market. Union and business interests clash again. The report emphasizes that the state is responsible for supporting regional areas (LRF 1986c).

The different opinions have not changed in recent years. The cooperatives tend to view rural rejuvenation as a responsibility of the farmers' union and the state. Tensions within the movement could be lessened were the state to assume greater responsibility. The general message is that the farmers' cooperatives are important for rural regions. They provide a "utility of regional policy" (*regionalpolitisk nytta*), another metaphor of the 1980s.

Much organizational effort is put in the development of new work tasks for farmers to keep them on their farms. The LRF has encouraged some ideas that can threaten the established cooperatives—for example, butcher and cheese shops on the farms. They are accepted as they further the union interests of farmers and make the countryside more attractive for urbanites. Some tasks further the basic principles of cooperation. Others fit neatly with postmaterialistic values. Agrotourism and biodynamically grown crops are good illustrations here.

REPRESENTING FARMERS IN CORPORATIST EXCHANGE

Opinion formation is important, but decisions on farming are made in political institutions and through political negotiating. The most prominent institutions have been the Riksdag and ministry of agriculture. It is not surprising that the LRF exerts much effort in its contacts with them. Contact with parliamentarians and government officials is commonly termed "lobbying." Swedish lobbying is done primarily by the established interest organizations and does not liken the American situation. Of all Swedish interest organizations, the LRF, LO, and SAF have been found to lobby the Riksdag most frequently (R&D 1988, no. 14; *Veckans affärer* 1988, no. 49).

The LRF concentrates its attention on agricultural or food policy issues. It contacts government ministers and officials, MPs, leaders of other interest organizations as well as journalists to create a better understanding for its view of the farm and food problem. Lobbying can range from regular contacts with particular policymakers to organized mass mailings to MPs on narrow issues. Sweden is interesting as much lobbying occurs in the newspapers: Leaders of interest organizations argue their standpoints in debate articles directed toward policymakers and the public. Another form of contact allows "grassroot lobbyists," individual members, to confront political leaders. For example, the organization created a setting where the minister of environment agreed to answer telephone calls from farmers shortly after she presented her bill on environmental policy in 1987.

The significant change for the LRF in regard to discussions with the state is that other ministries and their associated agencies are important contact partners. In particular, the ministries of environment and finance are involved now. The development reflects the integration of the food and agricultural sector into a larger policy and business setting.

As a consequence of this change, the LRF in 1989 reorganized its administration. The former separate departments for economic policy, price-setting, business, and industrial policies have been collapsed into one department that integrates the policy areas. Integration of policy areas seems to be a new trend in Sweden. A new unit that is a stock company is called LRF Consumer, which has the task of informing consumers about food products. The departments for information, member recruitment and training, the organizational newspaper, the research institute, and consultation services have been retained, though many of their structures and tasks have changed or expanded.

Lobbying is less important in Sweden than in the United States. The reason is the existence of an established system for the exchange of information in the food sector. It was termed the corporatist exchange in Chapter 5. One aspect of the exchange is the participation of interest organizations on the boards governing many state agencies. The LRF was represented on over 150 public bodies, including permanent, temporary and subordinant ones, in 1987. Its representation is concentrated in the areas of agricultural, business, regional, and food policy.

The representation of interest organizations on state institutions is both positive and negative for political democracy. It is positive for the organizations: it offers them the opportunity to discuss their views with other influential public actors. It is positive for society: the organizations are confronted with other viewpoints which frequently lead them to modify their standpoints and accept compromises. It is the responsibility of the leaders of the interest organizations to anchor the agreements within their memberships. Interest organizations assume the task as it is a prerequisite for their participation. All participants underscore that the organizations represent a more general rather than special group interest when they participate in the political process (SOU 1985:40; Micheletti 1985).

The conception of role is an important point. It helps explain the moderate position taken by Swedish interest organizations when compared to their counterparts in other countries. The LRF is a good example here. Also the conception of role is one defining characteristic of corporatism: "It integrates differing conceptions of group interest with vaguely but firmly held notions of the public interest" (Katzenstein 1985, 32).

The participation of interest organizations in state institutions is detrimental to democracy in two ways. The distinction between public and private responsibility becomes blurred, as discussed in Chapter 5. Moderation may not always be the strategy chosen by members. As shown in an earlier section, groups of members may appeal to their leaders to act more forcefully. Others may decide to leave the organization and join protest groups that are more steadfast in their proclamations.

The corporatist exchange complicates the choice that the leadership must make for the organization. Leaders must seek legitimacy from two sources. They need the support, loyalty, and response of the members in order to be able to lead the organization. Ideological control is important here. Solution of problems is the foremost task of leaders and involves change, chance, initiative, and decisionmaking. The trust of the membership is a very important asset when leaders are called upon to solve serious problems. A strong leadership knows that it must produce results, whether material or symbolic, and offer an interpretation of the situation that is acceptable to the membership (Wilensky 1967).

The current LRF leadership has been viewed as very adept in leading the organization through change. In particular, the president is singled out for this ability. Many of the metaphors used by the organization bear the mark of his leadership. They create a verbal environment for debate and have met with the support of the membership. The leadership has been successful in taking the initiative on change in policy settings. Farmers are seen as less defensive in the public debate, which has a positive effect on the membership as well.

The leadership must seek political legitimacy for the interest organization also. It is important that the organization is considered as legitimate by the political actors in the corporatist network. As shown in previous chapters, corporatism forces change on the interest organization. It is essential that leaders of interest organizations realize when their policies become outdated by events, and they must convince the members that this is the case. The ability to discuss change in the corporatist exchange is an attribute of the political legitimacy of the interest organization. It creates goodwill. The LRF has demonstrated recently that it has these qualities.

Yet it is unlikely that the LRF would undergo the process that a transformation to a postmaterialistic, deregulated agricultural policy requires were it not for pressure from actors outside the movement. Research has shown that the main function of policy is to help in the adaptation of institutions to changing orders. Scholars agree generally that flexibility and proclivity toward change is a characteristic of nations characterized as corporatist democracies (Katzenstein 1985).

A rather peaceful change in agricultural policy is occurring in Sweden now. High track decisionmaking on the general outlines of policy is combined with low track compromises over how responsibility for the transformation should be divided between the different actors. The two key ones are the state and the LRF. The important substantive issues are overproduction, the environment, and deregulation. Deregulation is discussed in Chapter 7.

Analysis of the treatment of the issues offers insight into the process of political bargaining. The issues of surplus production and the environment were handled differently. Also, their treatment explains the tendency of political scientists to analyze politics as a game complete with rules, sanctions, and rewards. A series of bargains and compromises occur on different levels within the framework of the corporatist exchange. The two issues show that the negotiating actors have different interests to protect. For the LRF, protection of grain farmers and the occupation of agriculture are major concerns. It does not want to bargain away all its political leverage by compromising itself unnecessarily. This may be extremely difficult to evaluate while negotiations take place.

In its initial stage, political bargaining is discussion on proposals to sound out the different involved actors. Sounding out the other participants can be done in a low or high track form. The two actors most engaged in this activity are the ministry of agriculture and the LRF. The LRF must also discuss and anchor proposals within its own heterogeneous membership.

The process does not always work smoothly when the protected interests of the participants are at stake. The corporatist exchange on solutions to the problem of surplus production were on the verge of breakdown. On one occasion, the tacit rules of corporatist etiquette were broken when the TCO questioned the democratic nature of the LRF organization. The LRF jeoparadized also the process by attacking both the government and the consumer representatives for "the politics of double talk" (*det dubbla budskapets politik*), a metaphor coined to express disappointment with what the LRF considers to be inconsistent politics and hypocrisy on the part of the ministry and the two labor unions.

This type of behavior can be expected when corporatist exchanges must deal with high track policymaking. The legitimacy of all participants is at stake. It reflects that leaders of interest organizations cannot only consider the well-being of the nation. They must cater to the needs of their members as well. The uncomfortable position for the LRF in the discussions on surplus production illustrates how difficult the corporatist exchange can be for an interest organization. It is worthy to note that policymaking discussions seldom remain on this verbal level in Sweden. The participants return to the low track to discuss issues more calmly and openly. Frequently it is the case that more dramatic speech is used as a rallying cry for support from the general public. Also, it tends to appease those members who criticize the leaders for compromising their interests unnecessarily.

Recently the LRF has been forced into a number of compromises that have disturbed groups of farmers. The treatment of the problem of surplus grain is one. The LRF has not had any success in convincing the government to finance the production of grain alcohol. The production of grain for alcohol is not very different from its cultivation for animal feed or flour. It is possible that the LRF favors this alternative as a way to appease large farmers and the grain growers' guild.

Policymaking is giving and taking. Lately, the LRF has given more than it has taken. Most disagreements over policy proposals have not been won by the farmers' movement. Adaptation to change characterizes the strategy of the LRF. The issue of surplus production is an example of defeat that is forcing change on the LRF. A related one is forest plantation, which is seen as a way to utilize the land. Foresty will change farming, the contours of the countryside, and the interests of farmers. Some agenda-setters fear that the open countryside—the Swedish heartland—will be endangered by it.

Yet the LRF is planning a strategy of adaptation. In many ways, the alternative seems a good one for the LRF. There is deficit rather than surplus production. The problem is the low membership density of the Swedish Forest Owners' Association, a situation which causes the LRF to emphasize the role of forest owners in the farmers' movement. Policy statements from the leadership stress the importance of forests that are owned by farmers. More attention is given to the combination of farming and forestry as a viable base for the members in the near future. In this way, the LRF is planning for a transformation of its membership and the countryside.

The treatment of the environmental issue of agriculture contrasts greatly with the problem solving of surplus production. The basic difference is the absence of a well-established network of corporatist exchange for the involved actors. Recent research has found that a corporatist channel for political exchange on the environment does not exist (Petersson 1989). Other differences between the two cases are apparent also. The issue of surplus production became an agenda issue because of its costs to the state, farmers, and consumers. The environmental issue grew in political importance due to public concern, which forced the state into action. Whereas the first issue was generated through the corporatist exchange network, the second one was initiated by forces outside the classic agricultural subgovernment.

The dramatic controversy over regulation of farmers close to the Bay of Laholm in southern Sweden will illustrate the difference. The case involved the problem of nitrogen running off from fields into the bay. An emergency plan that severely changed the situation for the farmers in the area was to be implemented. The local LRF union was highly involved in the farmers' protest against the plan. The protest proved to be very effective. The minister bowed to their criticism by accusing the local state authorities of misunderstanding the instructions. She said they had applied them too adamantly. Later she attempted to make amends to the farmers through the Center Party's expert on environment and agriculture and through direct and indirect contacts with the LRF. A compromise was gradually reached.

The tone of the debate has differed distinctively from the one discussed earlier. Mutual respect has been lacking. The problem of pollution has been overshadowed by the problem of trust and understanding. The environment minister has not always found that farmers willingly cooperate with her wishes. Farm leaders have tended to view her policy statements in terms of social democratic vote maximization (statement by LRF president in *Land* 1987, no. 17). They tend to disbelieve that the ministry understands the problems that agriculture faces in Sweden. The president has stated that farmers have taken a "beating" on the issue of the environment and has publically chided the minister for considering farmers "environmental pigs" (debate article by LRF president in *Skånska Dagbladet* 8 March 1987).

The environmental issue is a sensitive one for the farmers' movement. The greening of Sweden in the 1960s and early 1970s did not coincide with an acknowledgement of the role that agriculture plays in the pollution of nature. Agriculture and nature were often seen as synonymous. The Center Party did not stress the role of farmers as environmental polluters, and recent attempts to do so have been played down (Center Party 1987a, 1987b).

The environmental issue is also putting a strain on the LRF leadership, a situation that is not improved by the absence of a well-established corporatist exchange. The problem for the LRF is that it has only recently begun to prepare itself for the greening of agriculture. This may seem surprising, as the issue was given priority in the 1977 parliamentary decision on agricultural policy. Its delay as an item on the LRF agenda is explained by the difficulty of formulating a united members' interest on the issue.

The ability to adapt to change shows the strength of an interest organization. Adaptation is also a central characteristic of the Swedish political culture. It is not very surprising that the LRF has chosen this path for the future. If successful, it will be the first major interest organization to have undergone the transformation from industrialism to postindustrialism.

7

Farmers, Interest Organizations, and the State

GENERAL PROBLEMS OF COLLECTIVE ACTION

Collective action requires a sense of purposiveness. Interest organizations do not always need to reflect on their purposes, goals, or direction because collective action is not problematic. Representation of members is a routine task for the leadership. It knows what the interests of the membership are and how they should be articulated both within the organization and in external settings. It can be said that there is a congruency between the internal and external worlds of the organization. Also, the leadership receives the support of the members for its actions on their behalf. It is reelected, and members do not register complaints through protests. When purposeful action is an organizational routine there are neither severe problems inside nor in the outside sphere of the organization.

Yet purposeful action by an interest organization is not necessarily an easy task or given once and for all. The internal or external situation or both must be stable. At times, internal problems may arise while the external setting remains the same. Here the task for the leadership and membership is to improve organizational solidarity, that is, to unify the members in order to act together in a common interest to achieve a common purpose (Moe 1980). A change of leadership is perhaps one solution to this type of problem.

It may also be the case that the external setting changes, for example through a proposal for reform in public policy, while the organization remains unchanged internally. When this happens, the organization must decide how to react. Two obvious alternatives are mobilization of the membership for acceptance or rejection of the change.

The most difficult and complicated type of collective action for an interest organization is when internal and external change occur simultaneously. The changes in the two spheres may be even interrelated and dependent on each other. A decision to react to an external change may cause an internal change. A decision to solve an internal problem in a particular way may signify that the organization takes a stand on public policy issues. Collective action in this kind of uncertainty may signify that the organization finds it particularly difficult to know how to act. Also, it may find that its actions cause unintended consequences.

The presence of internal problems may impede the ability of the leadership to react to external change. It becomes difficult to know how to represent the interests of the membership, as internal conflict jeopardizes solidarity and the articulation of common member interests. The presence of external problems may well mean that internal problems cannot be given proper attention. Perhaps the organization decides to give priority to the external problems and ignore the internal ones. The leaders may pretend to have the support of the membership because they are pressured by external actors to take action. Inability to reach an internal compromise for the sake of political reform and development may injure the political legitimacy of the interest organization.

The nature of internal or external change or both may signify that the organization cannot decide on an issue until a policy decision has been anchored within the membership. It may even be the case that public policy alternatives have not been formulated in precise terms. Rather, it is generally understood that public policy will be reformed, trail balloons will be launched, and the policy reform will be formulated on the basis of reaction from the involved actors. The alternatives of choice for the interest organization are ill-defined and subject to change. Interest inarticulation—the inability of the organization to formulate how it will represent the interests of the membership on issues of importance—may be the only feasible alternative for the leadership (Micheletti 1987b).

This situation faces the LRF today. Change is occurring both internally and externally. Heterogenization of the union membership and adaptation of the farmers' cooperatives characterize the internal setting. The foremost changes in the external environment are the developments toward environmentalism, consumerism, and deregulation.

The uncertain situation means that it is not possible at this date to draw scholarly conclusions on the exact direction of policy and the LRF's stands on the issue. Rather this chapter brings together the different themes of analysis in the previous chapters. The first section discusses the problems of member solidarity while the second takes up the role of the LRF leadership in the reform process. The third section explains why reform is a pressing issue in Sweden and offers a brief comparison between the agricultural reforms of the 1960s and 1980s in terms of the problems they have created for the interest organizations of farmers. The final section focuses on the role of international forces in policy change.

INCENTIVES FOR MEMBERSHIP

People decide to join interest organizations for different reasons. The literature on membership in associations assumes generally that a person's decision to become a member is more or less rational and that a member will not remain in the organization if s/he believes that his or her interests are not represented adequately.

A central task of the leadership in this regard is organizational maintenance. Maintenance should not be narrowly seen as mere attempts by the leadership to stay in power or keep the organization in existence. It involves producing and sustaining the reasons for collective action, that is, "securing essential contributions of effort and resources from members, managing an effective system of communications, and helping formulate purposes" (Wilson 1973, 30)

A central task of organizational maintenance is the provision of incentives for membership to individuals. The incentives can be either tangible or intangible and aim to encourage individuals to join or remain as members. Four general types of incentives have been specified (Wilson 1973, ch. 3). Material incentives are tangible rewards that can be priced in monetary terms and allocated either to the entire membership (exclusive or selective benefits) or to particular members (individual benefits). Examples of both include fringe benefits, wage increases, and discounts on commodities and services.

The LRF provides exclusive as well as individual benefits to its members. Its provision of protection and selective incentives is discussed in Chapter 6. As pointed out there, nonmembers can receive certain benefits because the LRF identifies its services with farmers in general. It is noteworthy that the LRF does not put more emphasis on its task of collective bargaining, an example of an exclusive material benefit.

The goal of income equality has not been a primary incentive for the organization. This may seem rather puzzling since a majority of farmers state that profitmaking is an important concern and that they are worried about unfavorable agricultural prices (C. Johansson 1988). Statistics from Statistics Sweden (*Statistiska centralbyrån*) confirm the negative picture (SCB 1985c).

Different reasons can be given for the lower priority given to the goal of equality. A main one is the increasing importance of other gainful activities or additional income sources that supplement the farm income, a development not unique to Sweden. Roughly half of all Swedish farmers' income comes from outside sources. Additional income is more important for smaller farmers, but even the largest ones receive over half of their income from outside sources. The tendency is for increased supplements of the farm income. Other explanations include the accumulation of wealth by farmers, which compensates their lower incomes to some extent, as well as farm ownership, which has certain advantages over employment as a wage earner.

Another important reason concerns the heterogeneity of the farming population and the profile of the LRF membership. Certain farmers have better incomes than others, and they tend to be members of the LRF. Farmers with higher incomes are those from the plains of southern and middle Sweden (as opposed to farmers in the north and in the woodlands). They have also larger farms and specialize in crop cultivation (rather than animal husbandry). These farmers are dominant groups within the LRF. The majority of members are from the middle (about 46 percent) and southern (over 35 percent) parts of Sweden. The most productive Swedish farmland is in Skåne, which acounts for about 13 percent of the LRF farming membership (*LRF Aktuellt* 1988, no. 4). Also, the LRF is experiencing an upswing in member recruitment from the group of farmers with 100 or more hectares.

Four other explanations can be given for the lower priority for the goal of equality by the LRF. It does not seem to have been a pressing concern among the members. No large number of motions sent to the assembly by the county unions since the LRF was constituted in 1971 have dealt with the issue. Agricultural policymaking has affected how the farmers' organizations have reasoned about the goal as well. It will be recalled that the goal has generally been seen as a subsidiary policy goal or as a means to attain the main goals by social democratic governments.

A related explanation emphasizes the role played by the representatives of the consumers in the price-setting negotiations. The LRF can be said to have assumed a defensive position, for example in comparison to its Norwegian colleagues. The farmers' organizations of Norway negotiate directly with the ministry of agriculture and have pushed much more adamantly and successfully for clear formulation of the goal of equality and government commitments to attain it. Consumers are not incorporated into the negotiating system as in Sweden. It has been concluded that the presence of a countervailing power in the Swedish negotiations is an important reason for the moderation shown by the LRF (Steen 1981, 1988).

The encompassing nature of the LRF, which comprises both the union and cooperative branches of the farmers' movement, may have affected organizational priorities on the goal of equality also. A comparison with Norway shows that the union organizations of farmers give little concern to marketing questions. In contrast, the farmers' cooperatives in Sweden are giving increasing attention to consumer demands (Steen 1981).

The second kind of material benefits applies to individual members. These benefits vary in value and are given to specific individuals as compensation for their dedication to the organization. Study visits to other countries or parts of Sweden may be provided to certain members. A small farmers' organization (*Jordbrukarnas Riksförbund*) critical of the LRF reports that the LRF offers members who also belong to it study visits abroad in order to convince them to leave the organization.

Exclusive material benefits help recruit members and encourage them to remain in the organization. Individual benefits are incentives for members to participate more actively. Yet if the goals of the organization involve more long-term policy concerns, material incentives are of limited utility since they do not tend to further solidarity and organizational development. Solidary and purposive incentives are important in this regard.

Two kinds of solidary incentives can be specified. The value of both is measured in the esteem of the organization given by the members. The first is specific solidary incentives, or "intangible rewards arising out of the act of associating that can be given to, or withheld from, specific individuals" (Wilson 1973, 33f.). Possibly the best example is elected office. A recent study shows that about 15 percent of the LRF membership are elected officials (Petersson, Westholm, and Blomberg 1989). Other examples include honorary rewards and attention focused on individual members for their dedication.

Another kind of solidary incentive is generally seen as more crucial for organizational maintenance. It is collective solidary incentives, which promote the common interests of the group and pride in the organization. They discourage heterogenization of attitude. Perhaps one example is the availability of courses, discussions, and debates for members, which encourage member participation in organizational activities. A sense of belonging is a collective solidarity incentive. About 25 percent of the LRF membership participates actively in the organization, roughly the same level as in Swedish trade unions (Petersson, Westholm, and Blomberg 1989, 110).

The final incentive for membership is termed purposive. They are also intangible rewards. It is "the sense of satisfaction of having contributed to the attainment of a worthwhile cause," which often are assumed to be the official goals of an organization (Wilson 1973, 34). In contrast to solidary incentives, they are not dependent on praise or acknowledgement received by individual members. Self-interest, as narrowly construed, is not the motivation for member participation in and commitment to the organization.

Whereas solidary incentives aim solely at creating solidarity and oneness within an organization, purposive motives aim at fulfilling the basic objectives of the organization. While members may disagree on goals, strategy, and tactics, they support the general aim of the organization and want to develop and improve it. An active membership that meets to debate organizational issues openly and critically is important here.

All four types of incentives are crucial in times of organizational crisis. In different ways they create goodwill for the organization. They provide the leadership with alternative strategies of action; it can decide to emphasize one type of incentive over another to remedy a specific problem and cope with particular groups of members. It is also the case that different incentives appeal to different members and priority given by members to the incentives varies between groups and over time. Yet all incentives seem to be equally important when an organization experiences external and internal problems simultaneously.

This is the situation facing the LRF today. Many farmers express a general discontentment with their working situation. Weather and legislation ranked highest on their list of dissatisfaction in a study conducted in the early 1980s. The possibility to take a vacation, a material benefit offered by the LRF to farmers, as well as the state agricultural agencies, an area of corporatist exchange, are other issues of displeasure. A large proportion of farmers state also that they are displeased with the farmers' organizations, though a clear majority stated that they supported the agricultural policy formulated by the LRF (C. Johansson 1988).

LEADERSHIP IN CRISIS

Leaders can play a positive or negative role in organizational maintenance. Capable leadership is an important attribute of an organization in times of internal and external crisis. Successful leadership is probably the most essential incentive for membership because it combines both concrete demands and abstract expectations into purposeful goals and action for the organization as a whole. It symbolizes the strengths and weaknesses of the organization. When leadership is capable, it can convince most individual members that collective action requires them to compromise and possibly risk their specific self-interests—that influence over change requires a disciplined membership.

It is noteworthy that few researchers of interest organizations focus their attention on the general importance of organizational leadership. The actions and statements of leaders are recorded by scholars. Yet it seems that scholars have viewed leaders more as spokesmen for or heads of interest organizations. A distinction between "headship" and "leadership" is useful here. Leadership becomes headship when the functions performed by the leaders are routine, ceremonial, and do not require ingenuity and problem solving creativity (Gibb 1968).

Leadership is essential in organizations going through change. Two central characteristics for successful leadership in crisis and change are flexibility and awareness of alternatives and problems. Flexibility means that the leaders will not continue policies that have been dangerously outdated by events and other changes (Roos and Starke 1981). Awareness shapes organizational development and acts as a safeguard against passive and naive acceptance of internal or external change or both.

The task of leadership is to formulate problems and offer solutions to them in a way which takes into consideration both the demands of the membership and the structure of the external events and developments. In brief, leadership is innovation for the interest organization and relevant outside institutions and actors (Edelman 1988; Bryce 1921; Donaldson and Lorsch 1983).

Effective leadership on all levels of the organizational hierarchy has concerned the LRF. Chapter 6 takes up the topic of leadership within the LRF in terms of ideological control and heterogenization. The role of language, in particular metaphors, as a technique of the leadership was emphasized. The LRF is not the only Swedish interest organization challenged by change demanded by postmaterial developments. Yet it seems that it has been forced to cope with postmaterial demands much faster than most other large Swedish interest organizations.

The term "perestroika" has been used by journalists in their assessment of the general development of the LRF in the past year. Recently the LRF leadership began using the metaphor "change of track" (*spårbyte*) to summarize its willingness to discuss agricultural reform. Awareness of the problem is high among the leaders.

The LRF is to change track in four ways. The cooperatives will become more sensitive to market signals and demands of the consumers, and they will plan for the deregulation of agriculture offensively rather than defensively. More emphasis will be put on the environmental issue, and measures will be found to improve the situation of farms. Also, rural policy and the countryside will be given a higher priority, as change in agriculture should not be allowed to lead to rural decay. The LRF does not want regional and agricultural policy to become separate policy sectors. There are indications that the political parties, including the Center Party, are beginning to reason in this way. Whereas the LRF is concerned about agriculture as a vital industry in Sweden, the party emphasizes the issues of the cultural landscape and social structure of the country (interview with Olsson). Finally, the LRF will seek understanding and cooperation from all 6 parliamentary parties. Contact with the Center Party will no longer predominate (interview with Dockered).

The LRF is formulating its strategy for the future. The president explained the strategy by stating in a magazine interview that "we feel the wind of change that is blowing through society" (*Veckans Affärer* 1988, no. 49, my translation). The LRF newspaper quotes the executive officer responsible for the price-setting negotiations: "We must be open for the opinion against the present system of regulation" (*Land* 1988, no. 43). The leadership has shown awareness of alternative solutions to the problems. Detailed alternative solutions officially supported by the LRF have not been presented because they are still a subject for the corporatist exchange and internal organizational debate. Stability is sought by the leadership through a common solution to the internal and external problems.

The political culture of Sweden puts great emphasis on flexibility and compromise. It is not surprising that these values are emphasized by the LRF leadership. Yet it is interesting that it waited as long as it has before changing its policy.

The change can be explained by the realization that farmers are growing less important in the two channels where they were once very influential: electoral politics and the corporatist exchange. Their votes count less, for they are fewer in number. Their resources are less decisive, due to the development of a countervailing power and demands for change in agriculture.

The leadership realizes that a refusal to negotiate and an adamant defense of the present system of agricultural regulation will lead to lower political legitimacy and goodwill for the organization. Farm leaders in many other countries would probably be puzzled by the following statement of the LRF president: "It is not possible to be successful by only protesting, demonstrating and saying no" (*ATL* 1988, no. 48).

Swedish politics is fascinating for foreign scholars because influence can be more successfully exercised by sitting down in a meeting than by expressing articulated and adamant demands. Encompassing interest organizations will not always and necessarily oppose policy change and adjustment. Their standpoint will depend to a large degree on factors other than their strength in the issue network of the interest organization (Schonfield 1965, 199).

CORPORATISM AND FLEXIBILITY

Swedish political culture has been described as "an elite culture in which a highly pragmatic intellectual style, oriented toward the discovery of workable solutions to specific problems, structures a consensual approach to policy making" (Anton 1969, 99). Though written 20 years ago, the quotation offers a rather good characterization of the current political culture. Why this kind of political culture has developed in Sweden has interested researchers. There does not seem to be only one explanation to its development.

One convincing perspective emphasizes the significance of country size for political behavior and attitudes. Here the argument is that democratic corporatism tends in different degrees to characterize the small European states—Austria, Belgium, Denmark, Finland, the Netherlands, Norway, Sweden, and Switzerland—and that corporatism leads to flexibility in policymaking. Size is seen as affecting the characteristics of the political regimes of the countries, though it does not act as a determinant on political regime type. Small size and a viewpoint that all members of society are in the same boat seem to go hand in hand (Katzenstein 1985, 35).

The characteristics of the political regimes of the small European states are political centralism, state interventionism, the non-socialist bloc divided into political parties that compete to some degree with each other in elections, and proportional representation which is seen as encouraging a system of coalition or minority governments. The role played by social democratic parties deserves mention also. The political culture is characterized by conflict avoidance, compromise across the main social and political cleavages, and the participation of interest organizations in the governing process.

The division of the political right has allowed farmers to play a more important role politically and electorally than would be assumed by their mere numbers. This has occurred in the attempts by the Center and Liberal Parties to cooperate with each other in the 1960s, by the 3 parties of the right to create a non-socialist bloc in the 1970s and 1980s, and the coalition governments with the social democrats. Proportional representation and the importance of marginal voters and vote maximization in Swedish elections have also had favorable implications for farmers.

A second central characterization of the small European states is economic vulnerability. They are dependent on the import of modern investment goods and the export of more traditional consumer goods. Many scholars conclude that the economic openness of a country has a stronger effect on the development of corporatism than political regime characteristics (Katzenstein 1985, 81—87, 94—104). Corporatist interaction has been found to enhance predictability and acceptance of economic reforms. Rationalization of Swedish industry is one example. With the support of the Swedish state, industry rationalized and specialized its production in order to be able to compete effectively in the world market. The labor movement played a leading role in the process (LO 1961, 1966).

Farms were also rationalized. Their production became specialized as well. Yet the reasons for the reform were not economic competition, and the farmers' movement did not take the initiative for the reform. Reform was considered necessary for self-sufficiency and war preparedness. It was proposed by social democratic governments. Industry sought also manpower from the farms.

It is intriguing that the small European states show a stronger tendency than larger ones to develop two different economic sectors, one externally oriented and competitive, the other internally oriented and protected. The best example of the latter is agriculture. Swedish agricultural protection has recently been calculated (see Table 1.1) as one of the highest in the world among the OECD members and second only to Japan, a country characterized by a highly competitive industrial sector and highly protected agricultural sector (OECD 1988a, 1988b).

Now the social democrats want to change the situation. They have again taken the initiative to reform agriculture. After returning to office in 1982, they began developing a reform program for agriculture. Two state committees were appointed. Now a work group of parliamentarians (*parlamentarisk arbetsgrupp*) is tackling the problem. Except for the chairman, who was an undersecretary at the ministry of agriculture, all members come from the Standing Committee on Agriculture. The interest organizations have been called in as experts and seem to be playing a minor role in policy discussions. The work group was appointed by the government to study agricultural and food policy. The composition of the group differs from the state committee on agriculture appointed in 1960, which included representatives for the farmers as well as the consumers. It differs also from the 1982 commission on food policy, which included only representatives for the political parties among its members, some of whom were not MPs. It may be recalled that the 1986 commission on food policy was a one-man investigation.

The group is focusing its attention on border protectionism, regulation, as well as the goals of war preparedness and regional and environmental policy. An important issue is the formulation of policy means which will attain the goals of Swedish agricultural policy without created a number of unintended consequences. The same reason motivated the appointment of the 1962 commission on agricultural reform.

Now most actors generally agree that goal attainment has not been highly successful. The severity of the situation for agriculture internationally and the attempts in the 1970s to come to grips with high food prices through consumer subsidies have made goal attainment a higher priority. Its work is to be completed by 1990, and its conclusions are to form the basis for the government bill on agriculture reform.

Those involved in the work group as members or experts have been somewhat puzzled over their role and task. It was not known from the start, for example, whether the committee members would vote on proposals or ccould submit reservations, who would write the group report, and what members should do if they disagree with the report. Neither was it known whether the report would be sent out on a remiss referral (interview with Nyberg). One important member of the group acknowledged that he felt manipulated by the government and stated that it wants to break down opposition to the social democratic reform proposals (interview with Olsson).

The status of the parliamentary work group is interesting, for it is not a state committee (*statlig kommitté*) like the former investigations. It is an example of the general decline of the role played by the system of state committees, a unique Swedish tradition for policymaking and the corporatist exchange, that has been noted by political scientists and state officials. Nor is it listed among the groups, delegations, and councils included in an appendix to the report on state committees that is published on a yearly basis. It is a group within the ministry of agriculture which quite surprisingly includes only MPs among its members.

No formal instruction has been provided for its tasks. Rather, it has been given a general point of departure, as the government wants to ensure flexibility and openness in the discussions within the group. An important reason for the emphasis on flexibility is the inability to foresee how discussions will proceed within different relevant international organizations (GB 1988—89, no. 47, GB 1988—89, no. 100, Jo 1988a, 1988b).

INTERNATIONAL OPINION, DEREGULATION, AND CHANGE

Though seldom referred to in the agricultural reforms of the past, international thinking on the farm problem has played a role in Swedish policymaking. Compared to other Swedish ministries, the ministry of agriculture has quite a lot of contact with international and Nordic institutions, including the EC, EFTA, and Nordic Council (Petersson 1989, Table 9.3).

The issues of agricultural policy have been a very difficult area for the GATT since it was signed in 1947. Special restrictions on the principle of free trade have been accepted for agricultural products. Yet the GATT has come to realize that national support schemes must be a permanent issue for the negotiations on agricultural protection and that they can no longer be regarded as a purely national affair. Agricultural protection has been high on the agenda of the institution. Now there is a willingness to deal with the special treatment which has traditionally characterized agriculture and made it "a step-child of trade liberalization" (GATT 1984).

An agreement signed in spring 1989 is to lead to the deregulation of agriculture. It reflects concerns over a rise in the number of trade disputes and a perception that the rules have not always been effective in preventing the spread of policies which obstruct rather than facilitate structural adjustment. The contracting parties have pointed out that the industrial countries have used more aggressive anti-dumping and countervailing measures for protective purposes. The general understanding is that markets are increasingly interconnected and that the definition of trade policy can no longer be limited to actions taken at the border (GATT 1988).

The GATT is considered the most important international forum for change in agricultural trade by the government. One reason is the composition of the membership, which includes nations belonging to the European Community (EC), Organization of Economic Cooperation and Development (OECD), and European Free Trade Association (EFTA).

The impact given by the Swedish government to the GATT for domestic policymaking is impresssive. Also, it reflects the position of a small state in international organizations and the world market. The government has decided to allow the agreement to play a decisive role in the work of the parliamentary work group and the reform of agriculture. The basic reason for the decisions is the need to reform Swedish agricultural policy.

The plans to establish the single market in the EC by 1992 have also played a role in policymaking. Agricultural policy is considered generally as the one policy area within the EC where the ideas of interstate integration have been most successful. Agriculture is presently highly subsidized by the EC and accounts for over two-thirds of its annual budget. The agricultural policy goals accepted by the EC differ only slightly from the traditional Swedish ones (George 1985; Viklund 1987).

The costs of agricultural regulation, dissatisfaction with the policy program, as well as the problem of surplus production, the environment, and others led the EC Commission in 1985 to proclaim the reformation of agriculture. Many of the reform proposals are similar to the Swedish ones discussed now. Most important, the establishment of the single market by 1992 will necessitate a deregulation of agriculture by the member states (EC 1987).

Another prominent international body focusing attention on agricultural policy is the OECD. Its evaluation of the problem repeats the concerns of the other international institutions. The unintended consequences of agricultural subsidies and policy incentives are underscored (OECD 1987). A number of studies of specific countries were conducted as well. At least 8 countries have been investigated, among them Sweden. The report on Sweden reviews the attainment of the goals of Swedish agricultural policy. From the point of view of the OECD, it was concluded that "the price regulation policy...does not allow market signals directly or fully to influence the orientation of agricultural productions" (OECD 1988a, 17). Many of the measures implemented as agricultural policy were found to make consumer prices higher. Also, the goal of self-sufficiency was found to have been surpassed, and it was questioned whether the income objective had been achieved.

The study of Sweden, along with a later report comparing the studied OECD countries (OECD 1988b), has been used by proponents of reform in the debate on agricultural policy in Sweden. As Sweden has accepted the OECD goals for agriculture, which include a gradual deregulation, the assessment offered in the reports is direct criticism of state agricultural policy. Though the LRF has criticized the assessment, the reports are now considered a factual basis for policy discussions. They played a central role in the critical report on agricultural regulation presented by a group within the ministry of finance and are taken up in the parliamentary work group.

The issues of importance are the regulation and deregulation of agriculture, regional policy, and agricultural exceptionalism. They are three issues which seem to be permanent features of the political debate on agriculture. Also there are constellations of interests that represent the different viewpoints that can be expressed on the issues.

Chapters 2 and 4 stress the influence of what can be called the rural lobby in the Swedish political process. It consists of various groups whose interests have coincided on the preservation of the countryside and arable land. The interest organizations that represent regional interests belong to the lobby. Among them are the LRF.

Politicians from the different political parties can be classified as rural lobbyists as well. It is not uncommon for the MPs from a particular region to have the same opinion on issues of significance for their constituents and to express the opinion in parliamentary motions. It may also be recalled that votes from outlying regions in the elections to the Riksdag are important in different degrees for the political parties. Currently the farm vote is most important for the Center and Conservative Parties.

Influence is exercised in other ways. The rural lobby argues for its case in meetings with government officials and in the mass media. Perhaps journalists in rural and less populated areas can be seen as part of the lobby. It is noteworthy that the ministry of agriculture has more contacts with local than national radio stations than most other Swedish ministries (Petersson 1989, Table 8.7).

The rural myth is a type of peasant romanticism, which is a type of political romanticism. The rural issue is not seen as associated with politics and practical political judgments. It is understood as a question of moral value and survival (Schmidt 1986). It can prove difficult to argue convincingly for policy change when the policy area itself contains romantic elements. Agriculture is an area which has been highly romanticized in most countries. The Swedish love of the countryside has a long history. The lobby uses different methods to remind the citizenry and the political leaders of its rural past and of the significance of rural considerations in politics. One method discussed in Chapter 6 involved an expensive advertisement campaign for the largest cities that was financed by the LRF.

An investigation from the Study of Democracy and Power in Sweden (*Maktutredningen*) can be seen as a confirmation of the power of the rural myth. Residents of the countryside and farmers were found to believe that it is quite possible for them to exercise influence over their own situation though their political participation was lower than other groups. The term "quiet power" (*tystlåten makt*) was coined to characterize their perceptions. Other societal groups did not judge their possible influence as high, though they were more active and vocal (Petersson, Westholm, and Blomberg 1989, Table 6.13, 210).

Given the influence of the rural lobby on policymaking, it is not surprising that the social democrats have decided to stress international developments and conclusions reached by international associations as a major motivation for change in agricultural policy. Policy can be changed without confronting established domestic interests.

The government states that reform of agricultural policy is necessary for domestic considerations also. It points to the costs of regulation for the national budget and consumers. The ministry of agriculture believes that the goals of agricultural policy must be attained through other means than subsidies. Statistics from the SPK show that the agricultural products which are regulated have increased more in price when compared to nonsubsidized products. The conclusion is that agricultural regulation has not contributed to the attainment of the consumer goal of agricultural policy (SPK 1988).

This time the government and the Social Democratic Party seek to strike a different balance between the interests involved in agricultural policy. Its ambition signals that the role of farmers and their special interest organizations has changed considerably from the golden years of agriculture. A new balance has been struck between the goals of government, the organizational and membership interests of farmers, and the more diffuse interests of the market and consumers. Emphasis is put on the traditional issues involved in social democratic attempts to change the structure of industry: regulation policy, competition policy, and the contrasting interests of the sector (LO 1961, 50—60).

A statement on deregulation of the egg market by the LO's representative on the National Agricultural Marketing Board expresses pointedly the consequence in principle of the new orientation: "One should remember that just because there is a public goal does not necessarily mean that public measures are demanded for its attainment" (JN 1988b, my translation). A consistent and forceful application of this orientation will signify definite changes and the death of agricultural exceptionalism. One wonders whether it perhaps is the first clear indication of a new orientation for the Swedish welfare state.

References

PRIMARY SOURCES

GB (GOVERNMENT BILLS, KUNGL: MAJ:TS PROPOSITION, REGERINGENS PROPOSITION)

1947/No. 75. *Angående riktlinjerna för den framtida jordbrukspolitiken m. m.*

1959/No. 147. *Angående åtgärder i prisreglerande syfte på jordbrukets område m. m.*

1962/No. 193. *Angående omorganisation av statens jordbruksnämnds råd.*

1965/No. 100. *Angående ändrad organisation av rationaliseringsverksamheten på jordbrukets och trädgårdsnäringens områden.*

1967/No. 95. *Angående riktlinjer för jordbrukspolitiken m. m.*

1969/No. 107. *Angående reglering av priserna på jordbruksprodukter m.m.*

1975—76/No. 154. *Om reglering av priserna på jordbruksprodukter m.m.*

1977—78/No. 19. *Om nya riktlinjer för jordbrukspolitiken m.m.*

1983—84/No. 76. *Proposition om vissa livsmedelspolitiska frågor.*

1983—84/No. 84. *Om kooperationens kapitalförsörjningen m.m.*

1983—84/No. 182. *Om reglering av priser på jordbruksprodukter.*

1984—85/No. 166. *Om livsmedelspolitiken.*

1986—87/No. 95. *Totalaförsvarets fortsatta utveckling.*

1987—88/No. 66. *Sverige och den västeuropeiska integrationen.*

1987—88/No. 85. *Miljöpolitiken inför 90-talet.*

1987—88/No. 100. *Med förslag till statsbudget för budgetåret 1988/89. Bilaga 2. Gemensamma frågor.*

1987—88/No. 128. *Om miljöförbättrade åtgärder i jordbruket m.m.*

1987—88/No. 165. *Om reglering av priserna på jordbruksprodukter m.m.*

1988—89/No. 47. *Om vissa ekonomisk-politiska åtgärder, m.m.*

1988—89/No. 68. *Livsmedelskontroll.*

1988—89/No. 100. *Budgetpropositionen. Bilaga 11. Jordbruksdepartementet.*

RP (RIKSDAG MINUTES, RIKSDAGENS PROTOKOLL)

AK (Lower Chamber, Andra kammaren)
FK (Upper Chamber, Första Kammaren)
(Voting records are not included in the minutes but are available at the Riksdag Library.)

RM (RIKSDAG MOTIONS, MOTIONER)

1975—76/No. 1272. *Av Svante Lundkvist m. fl. Om vissa konsumentfrågor.*
1986—87/No. K212. *Av Bengt Westerberg m. fl. (fp). Korporatism i Sverige.*

JoU (STANDING COMMITTEE ON AGRICULTURE, JORDBRUKSUTSKOTTET)

1962/No. 34. *Utlåtande i anledning av kungl. maj:ts proposition angående omorganisation av statens jordbruksnämndens råd jämte i ämnet väckta motioner.*
1984—85/No. 33. *Betänkande om livsmedelspolitiken.*

OTHER PARLIAMENTARY PUBLICATIONS

Riksdag 1920, 1930, 1940, 1950, 1960, 1970, 1980. *Riksdagens årsbok.*
RRK 1979/No. 4, (Office of the Parliamentary Auditors, Riksdagens Revisorers Kansli). *Granskningsprema.*
RRK 1979—80/No. 17, *Riksdagens Revisorers förslag ang. jordbruksprisregleringen och regleringsföreningarna.*
Riksdag 1989. *Riksdagen 1988/89. Ledamöter och riksdagsorgan.* Riksdagens förvaltningskontor.
R&D. *Från Riksdag och Departement.* (Weekly newsletter on the government and parliament).

JO (MINISTRY OF AGRICULTURE, JORDBRUKSDEPARTEMENTET)

DsJo 1976: 1 (Report from the Ministry of Agriculture). *Statens jordbruksnämnds organisation m.m. Betänkande avgivit av utredningen om jordbruksnämndens organisation.*
DsJo 1983: 10. *1983 års livsmedelskommitté. Vissa jordbrukspolitiska förtursfrågor.*
DsJo 1986: 2. *Åtgärder för att minska spannmålsproduktionen på kort sikt. PM av spannmålsgruppen.*
DsJo 1987: 2. *Åtgärder för att minska spannmålsöverskottet och stimulera alternativ markanvändning. Rapport 3 från spannmålsgruppen.*
1921. *Betänkande rörande omorganisation av hushållningssällskapen med förslag till förordning angående ändrade allmänna grunder för hushållningssällskapens organisation m.m. samt hushållningssällskapens ekonomiska ställning och*

anslagsbehov. Angivet den 8 juni 1921 av särskilda av chefen för jordbruksdepartementet tillkallade sakkunniga. Stockholm: P.A. Norstedt & Söner.

1925. *Kungliga jordbruksdepartementet 1900—1925. Minnesskrift.*

1988a. *Förordnande i livsmedelspolitiska arbetsgruppen (Jo 198:B)*. Utdrag departementsprotokoll. 1988 12 12. Dnr. 2274/88. § 286.

1988b. *Tillkallande av en parlamentarisk arbetsgrupp för livsmedelspolitiska frågor.* Utdrag protokoll vid regeringssammanträde, 1 December 1988.

OTHER MINISTRIES

DsFi 1987: 4 (Report from the Ministry of Finance). *Vägar ut ur jordbruksprisregleringen. Bidrag till ett seminarium anordnat av expertgruppen för studier i offentlig ekonomi.*

Ds 1988: 54. *Alternativ i jordbrukspolitiken. Rapport till ESO.*

Glesbygdsdelegationen 10 November 1987 (Delegation for Development of Sparsely Populated Areas, Ministry of Industry). *Opinionsundersökning om attityder till landsbygden, m.m.* Hela Sverige Skal Leva.

PUBLIC AGENCIES

FOA 1988 (Försvarets forskningsanstalt, National Defense Research Institute). *Säkerhetspolitiska aspekter på livsmedelsförsörjningen.* FOA rapport C 10311—1.2.

JN 1988a. (Statens Jordbruksnämnd, National Agricultural Marketing Board) *Tvåprissystemet för mjölk.*

JN 1988b. *Konsekvenser av en avveckling av prisregleringen för ägg.*

LBS 1988. (Lantbruksstyrelsen, National Board of Agriculture) *Lantbruksnämnderna—en resurs för hela landsbygden. Handlingsprogram. Landsbygdsutveckling.*

NO 1981. (Näringsfrihetsombudsmannen, Competition Ombudsman) *Föreningsbyte och marknadsuppdelning—mjölk. Beslut.* 1981 06 12. Dnr. 343/80. Akt bil. 37.

NO 1988. Remiss reply. Dnr. 404/88. *Rapport (Ds Fi 1988:54) Alternativ i jordbrukspolitiken avgiven till Expertgruppen för studier i offentlig ekonomi (ESO).* 26 January.

RRV 1950—51, 1960—61, 1970—71, 1979—80. (Riksräkenskapsverket, later Riksrevisionsverket, National Auditing Board) *Budgetredovisning för budgetåret.*

RRV 1987—88. *Statsbudgetensutfall.*

RRV 1988. *Går det att förändra forskning? Exemplet Sveriges lantbruksuniversitet.* Revisionsrapport 19880903. Dnr. 1988:991.

SPK 1984: 2 (Statens Pris- och Kartellnämnd, National Price and Cartel Office; now Statens Pris- och Konkurrensverk, National Price and Competition Board). *Svinavel. Struktur och konkurrens.* SPKs utredningsserie.

SPK 1984. 20. *Nya livsmedelsprodukter. Fallstudien VispÄdel.*

SPK 1987:3. *Regleringar inom livsmedelsområdet. Kartläggning och analys.* SPKs utredningsserie.

SPK 1988. *Prisutvecklingen mellan januari 1983 och juli 1988 med fördelning på "reglerade" och "icke-reglerade" områden.* Photocopy.

Statskontoret 1983. (Agency for Administrative Development) *Vem skall betala jordbrukets rationalisering?* By Erik Lindgren, Statskontorets Småskrifter 9.

Statskontoret 1988. *Färre regler—bättre jordbruk. En debattskrift om jordbruksregleringen.* Rapport Nr. 9.

P&K. *Pris och konkurrens.* (Newsletter from the NO and MD).

SOU (STATE COMMITTEES AND COMMISSIONS, STATENS OFFENTLIGA UTREDNINGAR)

SOU 1964: 55. *Organisationen av rationaliseringsverksamheten m.m. på jordbrukets, skogsbruktets och trädgårdsnäringens område.*

SOU 1966: 30. *Den framtida jordbrukspolitiken. A: Bakgrunden.*

SOU 1966: 31. *Den framtida jordbrukspolitiken. B: Mål och medel.*

SOU 1966: 49. *Lantbruksnämndernas nya organisation m.m.*

SOU 1967: 7. *Den framtida jordbrukspolitiken. Remissyttranden.*

SOU 1977: 17. *Översyn av jordbrukspolitiken.*

SOU 1978: 9. *Ny konkurrensbegränsningslag. Betänkande av konkurrensutredning.*

SOU 1984: 86. *Jordbruks- och livsmedelspolitik. Huvudbetänkande av 1983 års livsmedelskommitté.*

SOU 1985: 40. *Regeringen, myndigheterna och myndigheternas ledning. Huvudbetänkande från verkledningskommittén.*

SOU 1987: 44. *Livsmedelspriser och livsmedelskvalitet. Betänkande av 1986 års livsmedelsutredning, LMU.*

INTERNATIONAL ORGANIZATIONS

EC 1986. (European Community) *Environment and Quality of Life. Agriculture and Environment: Management Agreements in Four Countries of the European Community.* Report EUR 10783.

EC 1987. "Environment and the CAP." *Green Europe.* Newsletter om the Common Agricultural Policy No. 3.

GATT 1984 (General Agreement on Trade and Tariffs) Committee on Trade in Agriculture. *Summary of Points Raised During the Meeting of the Committee Held on 26 and 27 September 1984. Notes by the Secretariat.* Photocopy.

GATT 1988. *International Trade 1987—88. Report from the GATT Secretariat.*

OECD 1987. (Organization for Economic Cooperation and Development) *National Policies and Agricultural Trade.*

OECD 1988a. *National Policies and Agricultural Trade. Country Study Sweden.*

OECD 1988b. *Agricultural Policies, Markets and Trade. Monitoring and Outlook.*

LRF'S PREDECESSORS

RLF and SL 1969. (Riksförbundet Landbygdens Folk and Svenska Lantbruksförbundet) *Lantbrukets organisationsutredning—en sammanfattning.* Stockholm: LTs Förlag.

RLF. *RLF information.*

LRF (LANTBRUKARNAS RIKSFÖRBUND)

Årsredovisning (Yearly reports from the CEO).
Land (Organizational newspaper).
Lantbruksåret. Fakta om lantbruket, lantbrukskooperationen och LRFs verksamhet (LRF's yearly reports).
LRF Information Aktuellt, later *LRF Aktuellt* (Newsletter to elected officials).
LRF Minutes 1971—88. Minutes from the General Annual Assemblies (Protokoll fört vid riksförbundstämmam and LRF-stämman). The minutes include the speeches by the LRF president and CEO, as well as member motions and board statements on these motions.
LRFs stadgar (Constitution).
Minutes from Meetings of the LRF Executive Board 1971—88.
Nya Perspektiv (Former organizational newspaper).
Perspektiv (Former organizational newspaper).
Representation (Representation on public and other boards).
1976. *LRFs finansieringsutredning.*
1979a. *LRFs ställning och funktion.* Photocopy.
1979b. *Lantbrukskooperativt handlingsprogram antaget av LRFs riksförbundsstämma 1979.*
1980. *Farmers' Cooperation in Sweden.* Stockholm: LTs Förlag. Together with the Swedish Cooperative Center.
1983. *Vi och dem i förening. Fakta och åsikter om LRFs fackliga och kooperativa roll.*
1985a. *Natur- och kulturminnesråd. Handlingsprogram antaget av LRFs riksförbundstämma 1985.*
1985b. *Nya Tag! Fakta och diskussion.*
1985c. *Nya Tag! Tillsammans.*
1986a. *Vår idé och framtiden. Ett studiematerial med nya perspektiv.* Stockholm: Lantbrukarnas Riksförbund and LTs Förlag.
1986b. *Lika pris för lika prestation. LRFs syn på prisdifferentiering.*
1986c. *Landsbygdspolitiskt handlingsprogram.* Photocopy.
1986d. *Nya Tag! LRF modellen.* Stockholm: LTs förlag.
1987a. *"EMIL-gruppens" rapport.* Photocopy.
1987b. *Lantbrukskooperationens idé och verklighet i ett framtid perspektiv. Diskussionsunderlag från "LIV-gruppen."*
1987c. *Prisa kvalitet. Miljöprogram för kvalitetsproduktion av svenska livsmedelsråvaror.*
1988a. *Bakgrundsbeskrivning.* Medlems- och utbildningsavdelningen. Dnr. 1033/88. Photocopy.
1988b. *Idé och framtid. Sammanfattning av slutrapport. Dynamisk dialog. Lantbrukskooperationens idé och framtid.*
1988c. *Strategiska mål för lantbrukarnas föreningsrörelse.*
1989. *Jordbruket i EG.* Stockholm: LTs förlag.

LRF MEMBER ASSOCIATIONS

ATL. (Annonsbladet till Tidningen Lantmännen, SLR organizational newspaper).

SLR 1955. (Svenska Lantmännens Riksförbund) *Svenska Lantmännens Riksförbund 1905—55*. Stockholm: LTs Förlag.
SS 1986. (Slakteriförbundet) *Medlemsnytta och förtroendemannaroll inom slakteriorganisationen*. Photocopy.

LRF SPEECHES

Dockered, Bo (LRF president) 1989. "Strategier, mål och visioner inför lantbrukets 90-tal." Vid LRF-Konferensen, 11 January.

LUI (AGRICULTURAL ECONOMICS RESEARCH INSTITUTE, LANTBRUKETS UTREDNINGSINSTITUT)

1987a. *Deltidslantbrukarna och LRF*. Photocopy.
1987b. *Trendrapport 1980—1986 över lantbrukarnas inställning till föreningarna*. Photocopy.
1988. *Frågor avseende möjligheterna att stärka LRF*. Photocopy.

POLITICAL PARTIES

Center Party 1987a. *Miljön, resursen och kvaliteten—jordbrukets framtid. Remissupplagan*. Photocopy.
Center Party 1987b. *Miljön, resursen och kvaliteten—jordbrukets framtid*. Stockholm: Center Party.
Westerberg, Bengt 1984. (Chairman of the Swedish Liberal Party). "Socialt ansvar utan socialism." Inledningsanförande av folkpartiets ordförande Bengt Westerberg på landsmötet den 23 november.

OTHER SWEDISH INTEREST ORGANIZATIONS AND POPULAR MOVEMENTS

LO 1961. (Landsorganisationen i Sverige, Confederation of Swedish Trade Unions) *Samordnad näringspolitik. Strukturutredningen*. Stockholm: Tiden.
LO 1966. *Fackföreningsrörelsen och den tekniska utvecklingen. Rapport från en arbetsgrupp till 1966 års LO-kongress*. Stockholm: Prisma.
LO 1985. *Livskraftigt. Forskare om jordbruk, mat och bioindustrin*.
TCO 1978. (Tjänstemännens Centralorganisation, Central Confederation of Professional Employees) *Konsumentfrågorna och tjänstemännen. Rapport från TCOs konsument- och jordbrukspolitiska arbetsgrupp*. Rapport Nr. 6 till TCO kongressen.
TCO 1981. *Livsmedelspolitiken och tjänstemännen*.
TCO 1987. *En förändrad livsmedelspolitik. Ett debattinlägg*.
TCO-tidningen (Newspaper to TCO's active members).

OTHER PRIMARY SOURCES

Arne K. Larsson AB 1988. *Kampanjmätning. Bönderna. Intervjuarbete i Stockholm och Göteborg den 27—29 juli 1988*. Photocopy.

Indevo Management Consultants 1989. *Översyn av jordbrukets regleringsföreningar*. Photocopy.
Relationskonsult 1987. *Opinionsbildning och information om svenskt lantbruk.*

INTERVIEWS

Bjärsdal, Jacob. Former official at the LRF, author of books and articles on the farmers' movement. 12 January 1987.
Bramsvik, Nils-Erik. Former official at the LRF, lobbyist for the LRF and the SS. 27 November 1986.
Carlsson, Bengt. Official employed at the KRAV. Telephone interview. 12 July 1988.
Dockered, Bo. President of the LRF. 7 April 1989.
Herland, Erik. The LRF representative in the policy group (*Spannmålsgruppen*) within the ministry of agriculture. Telephone interview. 24 May 1988.
Johansson, Anders V. Researcher, Department of Economics and Statistics, Swedish University of Agricultural Science. 8 February 1988.
Johansson, Lena. Assistant Undersecretary (*departementsråd*), ministry of agriculture, 27 January 1989.
Lanerstrand, Rune. President of Small Farmers in the West. Telephone interview. 12 July 1988.
Ljusberg, Jan-Erik. Chief Information Officer, National Price and Competition Board, Secretary of Livsmedelskommittén 1986. 19 January 1989.
Nyberg, Jan-Erik. The TCO expert on agricultural policy and representative on the Consumer Delegation, Chairman of the Consumer Delegation. 16 January 1989.
Olsson, Karl Erik. Vice Chairman of the Center Party, Chairman of the Riksdag's Standing Committee on Agriculture. 15 January 1988.
Ramberg, Hans-Olof. Head of the LRF Information Unit. 14 March 1989.
Sahlström, Sven. Division director in charge of monitoring agricultural regulation and competition, Competition Ombudsman. 30 January 1989.
Sylvan-Johansson, Margareta. Principal Administrative Director, Office of the Consumers' Delegation (*KDs sekretariat*). Telephone interview. 11 May 1988.
Waldenström, Cecilia. Researcher, Department of Economics and Statistics, Swedish University of Agricultural Science. 8 February 1988.

CORRESPONDENCE

Ahren, Steffan. President of the ARF. 12 August 1988.
Lundström, Agneta. Official at the Biologiska Odlare i Samverkan ek för. 20 December 1987.
Nyberg, Lennart. Official at the Jordbrukarnas Riksförbund. 11 February 1986.
Orjavik, Kari. Contact person for the Samodlarna Sverige. Spring 1988.

YEARBOOKS, STATISTICAL BOOKS, DICTIONARIES

När, Var, Hur. Stockholm: Bokförlaget Forum AB.
Nordisk familjebok 1913. Konversationslexikon och realencyklopedia. Stockholm: Nordisk familjeboksförlags aktiebolag.

SCB 1935 (Statistiska centralbyrån, Statistics Sweden). *Statistisk årsbok för Sverige.*
SCB 1970 Sveriges officiella statistik (SOS). *Allmänna valen 1970 Del 3. Riksdagsvalet. Specialundersökning.*
SCB 1969, 1971, 1978, 1980, 1985a. *Statsanställda. Del 2.*
SCB 1985b. *Offentliga sektorn. Utveckling och nyläge.*
SCB 1985c. SOS. *Den svenske bonden. Rapport nr. 46. En rapport om ekonomi, hälsa och arbetsmiljö.*

DAILY NEWSPAPERS

Dagens Nyheter
Skånska Dagbladet
Svenska Dagbladet

WEEKLY NEWS MAGAZINES

Affärsvärlden
Veckans Affärer

SECONDARY SOURCES

Åkerman, Gustaf. 1937. *Mjölkregleringen.* Stockholm: P.A. Norstedt & Sönner.
Alestalo, Matti, and Stein Kuhnle. 1984. *The Scandinavian Route. Economic, Social, and Political Developments in Denmark, Finland, Norway, and Sweden.* Helsinki: University of Helsinki. Research Group for Comparative Sociology Report No. 31.
Almond, Gabriel A., ed. 1974. *Comparative Politics Today. A World View.* Boston: Little, Brown and Co.
Als, Jeppe, and Bjarne Møgelhøj, eds. 1982. *Landbrugskooperationen i Norden. Medlemsaktivitet-medlemsindflydelse.* Esbjerg, Sydjysk Universitetsforlag.
Åmark, Karl. 1915. *Spannmålshandel och spannmålspolitik i Sverige 1719—1830.* Stockholm: Issac Marcus.
—. 1952. *Kristidspolitik och kristidshushållning i Sverige under och efter andra världskriget.* Vol.1 and 2. SOU 1952: 49—50.
Anderson, Åke. 1987. *Vårt jordbrukspolitiska system.* Uppsala: Swedish University of Agricultural Science, Department of Economics and Statistics, Report 284.
Andreae, Daniel. 1968. "Nationalromantik." *Emigrationen och det industriella genombrottet. Den svenska historien* 13: 192—94. Stockholm: Albert Bonniersförlag.
Andrlik, Erich. 1981. "The Farmers and the State: Agricultural Interests in West German Politics." *West European Politics* 4: 104—19.
Anton, Thomas J. 1969. "Policy-Making and Political Culture in Sweden." *Scandinavian Political Studies* 4: 88—102.

Atkinson, Michael M., and William D. Coleman. 1985. "Corporatism and Industrial Policy." In *Organized Interests and the State,* by Alan Cawson, ed. London: Sage.

Bäck, Kalle. 1984. *Bonde opposition och bonde inflytande under frihetstiden. Centralmakten och östgötaböndernas reaktioner i näringspolitiska frågor.* Stockholm: LTs förlag.

Bager, Torben, ed. 1985. *Landbrugskooperationen i Norden. Særpræg—Tillpassning.* Esbjerg: Sydjysk Universitetsforlag.

Ball, Alan, and Frances Millard. 1986. *Pressure Group Politics in Industrial Societies. A Comparative Introduction.* London: Macmillan.

Baron, P. J. 1978. "Why Co-Operation in Agricultural Marketing?" *Journal of Agricultural Economics 29*: 109—17.

Berger, Suzanne. 1969. "Corporative Organization: The Case of a French Rural Association." *Voluntary Associations NOMOS XI.* New York: Atherton.

—. 1972. *Peasants Against Politics. Rural Organization in Brittany 1911—1967.* Cambridge: Harvard University Press.

Bergström, Hans. 1987. *Rivstart? Från opposition till regering.* Stockholm: Tiden.

Berlin, Hjälmar, ed. 1936. *Bondeförbundets 25-årsjubileum.* Malmö: AB Skånska Dagbladets tryckeri.

von Beyme, Klaus. 1983. "Neo-Corporatism: A New Nut in an Old Shell?" *International Political Science Review* 4: 173—96.

Birgersson, Bengt Ove, et al. 1984. *Sverige efter 1900. En modern politisk historia.* 10th ed. Stockholm: BonniersFakta.

Bjärsdal, Jacob. 1983. *How We Organized the Farmers' Union. Experiences from the Pioneer Period of the Swedish Farmers' Union, RLF.* Stockholm: Swedish Cooperative Center and Federation of Swedish Farmers.

—. 1986. *Lantbrukskooperation och bondefack. Ett unikt samgående: förutsättningar, resultat och möjligheter.* Stockholm: LTs förlag.

Bolin, Olof. 1987. "Kooperationens dubbelroll och förnyelseproblemet." *Kooperativ årsbok:* 26—39.

Bolin, Olof, Per-Martin Meyerson, and Ingemar Ståhl. 1986. *Political Economy of the Food Sector. The Case of Sweden.* Stockholm: SNS förlag. Swedish ed., 1984.

Bondeson, Per Henrik. 1987. *Resultat av en undersökning om medlemsavgången inom LRF.* Stockholm: LRF. Photocopy.

Bonow, Mauritz. 1954. *Producent och konsument.* Föredrag hållet vid Sveriges lantbruksförbunds årsstämma. Stockholm: KFs bokförlag.

Bourdet, Yves. 1988. *Market Power and Consumer Welfare.* Uppsala: The Study of Power and Democracy in Sweden, English Series, Report No. 20. Photocopy.

Bramstång, Gunnar, ed. 1986. *Offentligrättsliga perspektiv på organisationernas verksamhet.* Skrifter från juridiska fakulteten i Uppsala, No. 8. Uppsala: Iustus förlag.

Browne, William P. 1982. "Farm Organizations and Agribusiness." In *Proceedings of the Academy of Political Science* 34: 3, Don F. Hadwiger and Ross B. Talbot (eds).

Bryce, James. 1921. *Modern Democracies.* Vol. 2. London: Macmillan.

Buksti, Jacob A. 1974. *Et enigt landbrug? Konflikt og samarbejde mellem landbrugets organisationer. En studie i landbrugets organisationsforhold med udgangspunkt i perioden 1957—61.* Aarhus: Universitetsforlaget.

Burns, Jim, et al., eds. 1983. *The Food Industry. Economics and Politics.* London: Heinemann.

Carlsson, Sten. 1953. *Lantmannapolitiken och industrialismen. Partigruppering och opinionsförskjutningar i svensk politik 1890—1902.* Stockholm: Lantbruksförbundets tidskrifts aktiebolag.

—. 1956. *Bonden i svensk historia.* Vol. 3. Stockholm: Lantbruksförbundets tidskrifts aktiebolag.

—. 1962. *Bonde, präst, ämbetsman. Svensk ståndscirkulation från 1680.* Stockholm: Prisma.

—. 1964. *Grupper och gestalter. Studie om individ och kollektiv i nordisk och europeisk historia.* Stockholm: Aldus/Bonniers.

—. 1966. "Den sociala omgrupperingen i Sverige efter 1866." *Samhälle och riksdag.* Vol. 1. Stockholm: Almqvist & Wiksell.

—. 1968a. "Den stora folkvandringen västerut." *Emigrationen och det industriella genombrottet. Den svenska historien* 13: 31—38. Stockholm: Albert Bonniers förlag.

—. 1968b. "Rationalisering inom jordbruket 1870—1920." *Emigrationen och det industriella genombrottet. Den svenska historien* 13: 132—37. Stockholm: Albert Bonniers förlag.

—. 1968c. *Yrken och samhällsgrupper.* Stockholm: Almqvist & Wiksell.

—. 1975. "Sweden in the 1760s." In *Sweden's Development from Poverty to Affluence,* by Steven Koblik, ed. Minneapolis: University of Minnesota Press.

Cawson, Alan. 1985. "Introduction: Varieties of Corporatism: The Importance of the Meso-Level of Interest Intermediation." In *Organized Interests and the State. See* Cawson, ed. 1985.

—, ed. 1985. *Organized Interests and the State. Studies in Meso-Corporatism.* London: Sage.

—. 1986. *Corporatism and Political Theory.* Oxford: Basil Blackwell.

Cigler, Allan J., and John Mark Hansen. 1983. "Group Formation Through Protest: The American Agriculture Movement." In *Interest Group Politics,* by Allan J. Cigler and Burdett A. Loomis, eds. Washington, D.C.: CQ Press.

Cigler, Allan J., and Burdett A. Loomis, eds. 1983. *Interest Group Politics.* Washington D.C.: CQ Press.

Clark, Burton R. 1972. "The Organizational Saga in Higher Education." *Administrative Science Quarterly* 17: 178—84.

Cox, Andrew. 1988. "The Old and New Testaments of Corporatism: Is It a Political Form or a Method of Policy-Making?" *Political Studies* 36: 294—308.

Crampton, John A. 1965. *The National Farmers' Union. Ideology of a Pressure Group.* Lincoln, Neb.: University of Nebraska Press.

Czarniawska-Joerges, Barbara. 1988. *Ideological Control in Non-Ideological Organizations.* New York: Praeger.

Czarniawska-Joerges, Barbara, and Bengt Jacobsson. 1989. "Budget in a Cold Climate." *Accounting, Organizations and Society* 14: 29—39.

Dahl, Erik. 1987. *Årsaker til oppslutning. En undersøkelse av faktorer som har betydning for medlemmenes oppslutning om Jönköpings Lantmän.* Ås: Norges Landbrukshogskole, Hovedoppgave vid Institutt for landbruksøkonomi.

—. 1988. *Förhållandet medlem-organisation. Jönköpings Lantmän.* Jönköping: Jönköpings Lantmän. Photocopy.

Dahl, Robert A., and Charles E. Lindblom. 1953. *Politics, Economics and Welfare.* New York: Harper and Brothers.

Dahl, Robert A., ed. 1966. *Political Oppositions in Western Democracies.* New Haven, Conn.: Yale University Press.

Daintith, Terence, ed. 1988. *Law as an Instrument of Economic Policy: Comparative and Critical Approaches.* Berlin: Walter de Gruyter.

Daudi, Philippe, and Richard Sotto. 1986. "European Cooperations in Trasition—The Metamorphos of Homo Cooperativus." *Scandinavian Journal of Management Studies* 3: 65—85.

Donaldson, Gordon, and Jay W. Lorsch. 1983. *Decision Making at the Top. The Shaping of Strategic Direction.* New York: Basic Books.

Dovring, F. 1965. *Problems of Manpower in Agriculture.* OECD Documentation in Food and Agriculture. Paris: OECD.

Drake, Lars. *Scenic Values of Agriculture.* Uppsala: Swedish University of Agricultural Sciences, Department of Economics and Statistics, Småskriftserien no. 4.

—. 1987. *Värdet av bevarat jordbruksslandskap. Resultat av intervjuundersökningar.* Uppsala: Swedish University of Agricultural Sciences, Department of Economics and Statistics, Report 289.

Edelman, Murray. 1988. *Constructing the Political Spectacle.* Chicago: University of Chicago Press.

Ekegren, Hjalmar, and Henrik Hagander. 1938. *Jordbrukets prisregleringar.* Stockholm: Lantbrukssällskapets tidskriftsaktiebolag.

Elander, Ingemar. 1978. *Det nödvändiga och det önskvärda. En studie av socialdemokratisk ideologi och regionalpolitik 1940—72.* Lund: Arkiv Avhandlingsserie.

Elder, Charles, and Roger W. Cobb. 1983. *The Political Use of Symbols.* New York: Longman.

Elder, Neil, and Rolf Gooderham. 1978. "The Centre Parties of Norway and Sweden." *Government and Opposition* 13: 218—35.

Elmér, Åke. 1975. *Från fattigsverige till välfärdsstaten. Sociala förhållanden och socialpolitik i Sverige under nittonhundratalet.* 6th ed. Stockholm: Aldus/Bonniers.

Elvander, Nils. 1965. Review of *Föreningsdemokrati och förtroendemannakår inom jordbrukets föreningsrörelse i början av 1950-talet,* by Gunnar Hellström. *Statsvetenskaplig tidskrift:* 114—19.

—. 1972. *Intresseorganisationerna i dagens Sverige.* 2nd ed. Lund: CWK Gleerup.

Enochsson, Jorma, and Roland Petersson. 1973. *Gunnar Hedlund.* Stockholm: Norstedts.

Erlander, Tage. 1973. *1940—1948.* Stockholm: Tiden.

—. 1974. *1949—1954.* Stockholm: Tiden.

Esping-Andersen, Gøsta. 1985. *Politics Against Markets. The Social Democratic Road to Power.* Princeton, N.J.: Princeton University Press.

Finger, J. M., Keith H. Hall, and Douglas R. Nelson. 1982. "The Political Economy of Administered Protection." *American Economic Review* 72: 452—66.

Fite, Gilbert C. 1981. *American Farmers. The New Minority.* Bloomington, Ind.: Indiana University Press.

Frängsmyr, Tore. 1977. *Vägröjare i kulturdebatt.* Stockholm: LTs förlag.

Friberg, Lennart. 1973. *Styre i kristid. Studier i krisförvaltningens organisation och struktur 1939—1945.* Stockholm: Allmänna förlaget.

Frykman, Jonas, and Orvar Löfgren. 1987. *Culture Builders. A Historical Anthropology of Middle Class Life*. Translated by Alan Crozier. New Brunswick, N.J.: Rutgers University Press.

Galbraith, John Kenneth. 1952. *American Capitalism. The Concept of Countervailing Power*. Boston: Houghton Mifflin.

Galston, William A. 1985. *A Tough Row to Hoe: The 1985 Farm Bill and Beyond*. New York: Hamilton Press.

Gardner, Bruce L. 1983. *The Governing of Agriculture*. Lawrence, Kan.: University of Kansas Press.

Gåsdal, Odd. 1985. "Konflikt i landbruket—splittelse i organisasjonerne?" *Landbruksøkonomisk forum* 2: 4—11.

Gellerman, Olle. 1958. *Staten och jordbruket 1867—1918*. Stockholm: Almqvist & Wiksell.

George, Aurelia D. 1981. "The Japanese Farm Lobby and Agricultural Policy-Making." *Pacific Affairs* 53: 409—30.

George, Stephen. 1985. *Politics and Policy in the European Community*. Oxford: Claredon Press.

Gibb, Cecil A. 1968. "Leadership: Psychological Aspects." *International Encyclopedia of the Social Sciences*. Vol. 9. New York: Macmillan and Free Press.

Grant, Wyn. 1985. "Corporatism and the Public-Private Distinction." In *State and Market*, by Jan-Erik Lane, ed. London: Sage.

Grant, Wyn, and Jane Sargent. 1987. *Business and Politics in Britain*. Basingstoke: Macmillan.

Grant, Wyn, ed. 1987. *Business Interests, Organizational Development and Private Interest Government. An International Comparative Study of the Food Processing Industry*. Berlin: Walter de Gruyter.

Grant, Wyn, and William Coleman. 1987. "Conclusions." In *Business Interests, Organizational Development and Private Government*, by Wyn Grant, ed. Berlin: Walter de Gruyter.

Grosskopf, Göran. 1981. *Ekonomiska föreningar*. Lund: Studentlitteratur.

Gulbrandsen, Odd, and Assar Lindbeck. 1973. *The Economics of the Agricultural Sector*. Stockholm: Almqvist & Wiksell.

Gusfield, Joseph. 1963. *Symbolic Crusade*. Urbana, Ill.: University of Illinois Press.

Gustafsson, Agne. 1986. "Rise and Decline of Nations: Sweden." *Scandinavian Political Studies* 9: 35—50.

Gustafsson, Bo. 1974. "Perspektiv på den offentliga sektorn under 1930-talet." In *Kriser och krispolitik in Norden under mellankrigstiden*. Stockholm: Almqvist & Wiksell.

Hadwiger, Don F., and William P. Browne. 1978. *The New Politics of Food*. Lexington, Mass.: Lexington Books.

Hadwiger, Don F., and Ross B. Talbot, eds. 1982. "Food Policy and Farm Programs." *Proceedings of the Academy of Political Science* 34: 3.

Hammerich, Kai. 1977. *Kompromissernas koalition. Person- och matkspelet kring regeringen Fälldin*. Stockholm: Rabén & Sjögren.

Hansen, John Mark. 1987. "The Ever-Decreasing Grandstand: Constraint and Change of an Agricultural Policy Network, 1949—1980." Paper for the Annual Meeting of the American Political Science Association, Chicago, Ill., 3—6 September.

Håstad, Elis, et al. 1950. *"Gallup" och den svenska väljarkåren. Några studier om opinionsmätningar.* Stockholm: Almqvist & Wiksell.
Heckscher, Gunnar. 1951. *Staten och organisationerna.* 2nd ed. Stockholm: Kooperativa Förbundets bokförlag.
Heclo, Hugh. 1979. "Issue Networks and the Executive Establishment." In *The New American Political System,* by Anthony King, ed. Washington, D. C.: American Enterprise Institute.
Hedlund, Stefan, and Mats Lundahl. 1985a. *Beredskap eller protektionism? En studie av beredskapsmålet i svensk jordbrukspolitik.* Stockholm: Liber.
—. 1985b. "Är vår jordbruksberedskap god?" *Ekonomisk debatt*: 397—404.
von Heland, Erik. 1969. *Optimismens och besvikelsens år 1922—1952.* Stockholm: LTs förlag.
Hellström, Gunnar. 1958. *Slakterinäringen inför 30-talets omorganisation.* Stockholm: LTs förlag.
—. 1964. *Föreningsdemokrati och förtroendemannakår inom jordbrukets föreningsrörelse i början av 1950-talet.* Stockholm: LTs förlag.
—. 1976. *Jordbrukspolitik i industrisamhället med tygndpunkt på 1920- och 30-talen.* Stockholm: LTs förlag.
Herlitz, Nils. 1936. *Grunddragen av det svenska statsskickets historia.* 2nd ed. Stockholm: P.A. Norstedts & Sönner.
Holmberg, Sören. 1974. *'Riksdagen representerar svenska folket'. Empiriska studier i representativ demokrati.* Lund: Studentlitteratur.
Holmberg, Sören, and Mikael Gilljam. 1987. *Väljare och val i Sverige.* Stockholm: Bonniers.
Holmberg, Sören, and Peter Esaiasson. 1988. *De folkvalda. En bok om riksdagsledamöterna och den representativa demokratin i Sverige.* Stockholm: Bonniers.
Holmström, Barry. 1983. *Äganderätt och brukningsansvar. Idéer och intresse i svensk jordpolitik.* Stockholm: Almqvist & Wiksell Internationell.
Holmström, Barry. 1987. "Marklagstiftningen som intressekamp." In *Festskrift till Professor Skytteanus Carl Arvid Hessler,* Leif Lewin, ed. Stockholm: Almqvist & Wiksell Internationell.
Horsfield, Jim. 1986. "Jim Horsfield on Our Farms as a National Park. We Need Farmers to Maintain It." *Choices* 3: 38.
Ingulfson, Hans, and Rolf Hagman. 1950. "De svenska partiernas sociala och andra grundvalar." In *"Gallup" och den svenska väljarkåren. See* Håstad et al.
Isberg, Magnus, et al. 1974. *Partierna inför väljarna. Svensk valpropaganda 1960—1966.* Stockholm: Allmänna Förlag.
Ivarsson, Rune. 1977. *Jordförvärvslagen, socialdemokratin och bönderna. En studie över jordförvärvslagstiftningen i svensk politik 1945—1965.* Malmö: CWK Gleerup.
Jacobsson, Bengt. 1984. *Hur styrs förvaltningen? Myt, och verklighet kring departementens styrning av ämbetsverken.* Lund: Studentlitteratur.
Johanssen, Knut, et al. 1987. *Heroer på offentlighetens scen. Politiker och publicister i Sverige 1809—1914.* Stockholm: Tiden.
Johansson, Anders W. 1987. *Utveckling och marknadsföring av konsulttjänster till lantbrukare.* Uppsala: Swedish University of Agricultural Sciences, Department of Economics and Statistics, Report 285.
Johansson, Curt R. 1988. Unpublished statistics from the project "Psykologiska

och sociala aspekter på lantbrukets arbetsmiljö." Lund: Department of Psychology, University of Lund.

Johansson, Inge. 1985. *För folket och genom folket.* Stockholm: Liber.

Johansson, Tore. 1978. *Lantbrukarnas föreningsrörelse. Framväxt och utveckling.* Stockholm: LTF.

—. 1987. "Lantbrukskooperationen idéerna och 80-talet." *Kooperativ årsbok:* 51—68.

Johansson, Tore, and Per Thullberg. 1979. *Samverkan gav styrkan. Lantbrukarnas föreningsrörelse 1929—1979.* Stockholm: LTs förlag.

Johnson, D. Gale. 1973. *World Agriculture in Disarray.* London: Macmillan.

Jonasson, Gustaf. 1972. "På väg mot en koalition. Förhandlingarna mellan socialdemokrater och bondeförbundare efter 1948 års val." *Historisk tidskrift:* 353—408.

Jonnergård, Gustaf. 1985a. *Med Gunnar Hedlund i politiken.* Stockholm: LTs förlag.

—. 1985b. *Så blev det centerpartiet. Bondeförbunds- och centeridéerna från fyrtiotalet fram till 1960.* Stockholm: LTs förlag.

Julen, Ann Sofi. 1982. *1959 års jordbrukspolitiska beslut.* Stockholm: University of Stockholm, Department of History. Photocopy.

Katzenstein, Peter J. 1985. *Small States in World Markets. Industrial Policy in Europe.* Ithaca, N. Y.: Cornell University Press.

Keeler, John T. S. 1987. *The Politics of Neocorporatism in France. Farmers, The State, and Agricultural Policy-Making in the Fifth Republic.* Oxford: Oxford University Press.

Kelman, Steven. 1981. *Regulating America, Regulating Sweden: A Comparative Study of Occupational Safety and Health Policy.* Cambridge, Mass.: MIT Press.

Kempe, Erik. 1923. *Studier angående de svenska hushållningssällskapen. Med särskild hänsyn till ett årshundrades förvaltningspolitik på lantbrukets område.* Uppsala: Almqvist & Wiksell.

Kihlberg, Leif. 1922. *Den svenska ministären under ståndsriksdag och tvåkammarsystem intill 1905 års totala ministerskiftet.* Uppsala: Almqvist & Wiksell.

King, Anthony, ed. 1979. *The New American Political System.* 2nd ed. Washington, D. C.: American Enterprise Institute for Public Policy.

Kingdon, John W. 1984. *Agendas, Alternatives, and Public Policies.* Boston: Little, Brown and Co.

Koblik, Steven, ed. 1975. *Sweden's Development from Poverty to Affluence, 1750—1970.* Minneapolis: University of Minnesota Press.

Koning Niek. 1983. "Family Farms and Industrial Capitalism." *The Netherlands' Journal of Sociology* 19: 29—46.

Korpi, Valter. 1978. *The Working Class in Welfare Capitalism: Work, Union and Politics in Sweden.* London: Routledge & Kegan Paul.

Kriser och krispolitik i Norden under mellankrigstiden. Nordiska historikermötet i Uppsala 1974. 1974. Stockholm: Almqvist & Wiksell.

Kristersson, Helge. 1954. "Samhällsekonomien och organisationsväsendet." *Samhällsdebatten,* No. 5.

Kylebäck, Hugo. 1984. *Konsument- och lantbrukskooperation i Sverige. Utveckling, samarbets- och konkurrensförhållanden före andra världskriget.* Stockholm: Kooperativa institutet.

Lane, Jan-Erik, ed. 1985. *State and Market. The Politics of the Public and the Private*. London: Sage.

Larsson, Hans Albin. 1980. *Partireformationen—Från bondeförbundet till centerparti*. Lund: GWK Gleerup.

Laswell, Harold D. 1965. *Language of Politics*. Cambridge, Mass.: MIT Press.

Laumann, Edward O., and David Knoke. 1987. *The Organizational State. Social Choice in National Policy Domains*. Madison, Wisc.: University of Wisconsin Press.

Laux, William E. 1968. "Agricultural Interest Groups in Danish Politics: An Examination of Group Frustration Amidst Political Stability." *Western Political Quarterly* 21: 436—55.

Lehmbruch, Gerhard. 1987. "Comparative Political Economy of Neo-Corporatism: Inter-Organizational and Institutional Logics." Paper for the European Consortium of Political Science, Amsterdam, Holland, 10—15 April.

Lewin, Leif. 1967. *Planhushållningsdebatten*. Uppsala: Almqvist & Wiksell.

—, ed. 1987. *Festskrift till Professor Skytteanus Carl Arvid Hessler. Utgiven till 80-årsdagen den 10 februari 1987 av hans lärjungar*. Stockholm: Almqvist & Wiksell International.

Lipset, Seymour Martin. 1971. *Agrarian Socialism. The Cooperative Commonwealth Federation in Saskatchewan. A Study in Political Sociology*. 2nd ed. Berkeley: University of California Press.

Lipset, Seymour Martin, and Stein Rokkan, eds. 1977. *Party Systems and Voter Alignments*. New York: Free Press.

Lönnroth, Erik. 1989. "Regional and National Representation. The Problems of Communication in Older Times." In *The Swedish Riksdag in an International Perspective*, by Nils Stjernqvist, ed. Stockholm: Bank of Sweden Tercentenary Foundation.

Loomis, Burdett A., and Allan J. Cigler. 1983. "Introduction: The Changing Nature of Interest Group Politics." In *Interest Group Politics*. *See* Cigler and Loomis, eds.

Lundholm, Kjell G., et al. 1982. *Kooperativ företagsamhet. Om ideologi och kooperativ företagsamhet*. Stockholm: LTs förlag.

Lundholm, Kjell G. 1986. *Lantbrukskooperationen inför 90-talet*. Stockholm: Liber.

Lundqvist, Lennart. 1971. *Miljövårdsförvaltning och politisk struktur*. Lund: Prisma/Verdandidebatt.

—. 1980. *The Hare and the Tortoise. Clean Air Policies in the United States and Sweden*. Ann Arbor, Mich.: University of Michigan Press.

Manchester, Alden. 1983. *The Public Role in the Dairy Economy*. Boulder, Colo.: Westview Press.

Månsson, Olle. 1976. *Industriell beredskap. Om ekonomisk försvarsplanering inför andra världskriget*. Stockholm: Liber.

Marcusson, Lena. 1986. "Organisationernas roll i det offentliga styrningssystemet enligt svensk rätt." In *Offentligrättsliga perspektiv på organisationernas verksamhet*. *See* Bramstång, ed.

Mather, Lynn, and Barbara Yngvesson. 1980—81. "Language, Audience, and the Transformation of Disputes. " *Law and Society Review* 15: 775—821.

McFarland, Andrew S. 1987a. "Interest Groups and Theories of Power in America." *British Journal of Political Science* 17: 129—47.

—. 1987b. "Sources of Countervailing Power in America: Contributions from Recent Interest Group Theory." Paper for the Annual Meetings of the American Political Science Association, Chicago, Ill., 3—6 September.

McInerney, John. 1983. "A Synoptic View of Policy-Making in the Food Sector." In *The Food Industry*. *See* Burns et al.

Merelman, Richard M. 1966. "Learning and Legitimacy." *American Political Science Review* 60: 548—61.

—. 1969. "The Dramaturgy of Politics." *Sociological Quarterly* 10: 216—41.

Micheletti, Michele. 1985. *Organizing Interest and Organized Protest. Difficulties of Member Representation for the Swedish Central Organization of Salaried Employees (TCO)*. Stockholm: University of Stockholm, Stockholm Studies in Politics, 29.

—. 1987a. "Organization and Representation of Farmers' Interests in Sweden." *Sociologia Ruralis* 27: 198—212.

—. 1987b. "Toward Interest Inarticulation. A Major Consequence of Corporatism for Interest Organizations." Paper for the Annual Meeting of the American Political Science Association, Chicago, Ill., 3—6 September.

—. 1988. "De svenska intresseorganisationerna—igår, i dag och i morgon." *Statsvetenskaplig tidskrift:* 41—54.

—. 1989. "The Swedish General Election of 1988." *Electoral Studies* 8: 169—74.

Misgeld, Klaus, Karl Molin, and Klas Åmark, eds. 1988. *Socialdemokratins samhälle. SAP och Sverige under 100 år.* Stockholm: Tiden.

Moe, Terry M. 1980. *The Organization of Interests. Incentives and the Dynamics of Political Interest Groups*. London: University of Chicago Press.

Möller, Tommy. 1986. *Borgerlig samverkan*. Uppsala: Diskurs.

Municio, Ingegerd. 1987. *Från lag till bruk. Hemspråksreformens genomförande.* Stockholm: University of Stockholm, Stockholm Studies in Politics, Nr. 31.

Nadel, Mark and David Vogel. 1977. "Who is a Consumer? An Analysis of the Politics of Consumer Conflict." *American Political Quarterly* 5: 27—56.

Näsgård, Bernhard. 1936. "Bondeförbundarna i Riksdagen." In *Bondeförbundets 25-årsjubileum*. *See* Berlin, ed.

Nielsen, Klaus, and Ove K. Pedersen. 1988. "The Negotiated Economy: Ideal and History." *Scandinavian Political Studies* 11: 79—101.

Niléhen, Lars. 1984. "Inför vägskälet. Nyare forskning kring jordbruket, bondeförbundet och centerpartiet 1945—1976." *Historisk tidskrift:* 422—31.

Nilsson, Jerker. 1986. *Den kooperativa verksamhetsformen*. Lund: Studentlitteratur.

Nilsson, Kjell-Arne, and Carl Utterström. 1988. *Lantbruket och marknaden. Varornas väg från lantbrukarna till konsumenten*. Stockholm: LTs förlag.

Nitsch, O. Ulrich. 1972. "Farmers' Problems in Practical Adoption and the Role of the Agricultural Extension Agent." Master's thesis, University of Wisconsin at Madison.

—. 1977. "Reinforcement of Official Goals in the Swedish National Board of Agriculture and District Agricultural Boards." Paper for the Department of Urban and Regional Planning Seminar.

—. 1982. *Farmers' Perception of and Preferences Concerning Agricultural Extension Programs*. Uppsala: Swedish University of Agricultural Sciences, Report No. 195.

Nyman, Olle. 1944. *Krisuppgörelsen mellan socialdemokraterna och bondeförbun-*

det 1933. Uppsala: Almqvist & Wiksell.
Nystrom, Paul C., and William H. Starbuch, eds. 1981. *Handbook of Organizational Design*. Vol. 1. *Adapting Organizations to their Environments*. Oxford: Oxford University Press.
Odhner, Claes-Erik. 1988. "Arbetare och bönder formar den svenska modellen." In *Socialdemokratins samhälle. See* Misgeld, Molin, and Åmark, eds.
Offe, Claus. 1985. *Disorganized Capitalism. Contemporary Transformations of Work and Politics*. Edited by John Keane. Cambridge, Mass.: MIT Press.
Olivecrona, Gustaf. 1968. *Hur väljarna vanns. Ett politiskt reportage*. Stockholm: Wahlström & Widstrand.
Olsen, Johan P. 1987. "The Modernization of Public Administration in the Nordic Countries. Some Research Questions." Paper for the Third Finnish Conference of Administrative Studies, the Finnish Society for Administrative Research, Helsinki, Finland, 19 November.
—. 1988. *Towards a New Public-Sector Revolution?* Uppsala: The Study of Power and Democracy in Sweden, English Series, Report No. 14. Photocopy.
Olson, Mancur. 1975. *The Logic of Collective Action. Public Goods and the Theory of Groups*. 5th ed. Cambridge, Mass.: University of Harvard Press.
—. 1982. *The Rise and Decline of Nations. Economic Growth, Stagflation and Social Rigidities*. London: Yale University Press.
Olsson, Helmer. 1945. *Jordbrukspolitiken under krisen*. Stockholm: Kooperativa Förbundets bokförlag.
Ovesen, Thorkild. 1958. "Swedish Agricultural Policy and Agricultural Production from 1930 to 1949." *Economy and History* 1: 43—63.
Pedersen, Ove. 1988. "Den sonderdelte stat." *Statsvetenskaplig tidskrift:* 259—78.
Persson, Knut. 1983. *Om LRFs tillkomst och verksamheter det första decenniet*. Uppsala: University of Agricultural Sciences, Department of Economics and Statistics, No. 58.
Peterson, Martin. 1985. "'The Family Farm': A Review of a Central Concept in Western European Agricultural Policies." *Scandinavian Journal of History* 11: 265—95.
Petersson, Olof. 1981. "Partier och väljare." *Känn Ditt Land*, No. 11. Stockholm: Svenska turistföreningen.
—. 1989. *Maktens nätverk*. Stockholm: Carlssons.
Petersson, Olof, Anders Westholm, and Göran Blomberg. 1989. *Medborgarnas makt*. Stockholm: Carlssons.
Qvist, Per Olov. 1986. *Jorden är vår arvedel. Landsbygden i svensk spelfilm 1940—1959*. Uppsala: Filmhäftet.
Rabinowicz, Ewa, Ingemar Haraldsson, and Olof Bolin. 1986. "The Evolution of a Regulation System in Agriculture. The Swedish Case." *Food Policy*, November: 223—33.
Reuterskiöld, C.A. 1936. "Hushållningssällskapen och förvaltningen." *Statsvetenskaplig tidskrift:* 1—10.
Ringborg, Gunnar. 1940. *Redogörelsen för jordbrukarnas ekonomiska föreningsrörelse*. Bilaga 2. SOU 1940: 35. *Organiserad samverkan inom svenskt näringsliv. Betänkande angivet av 1936 års näringsorganisationssakkunniga*.
Roald, Knut Halvard. 1977. *Mjølk som våpen. Aksjonsrørsla i norsk jordbruk 1945—1947*. Oslo: University of Oslo. Photocopy.
Røkholt, Per Ove. 1986. "Per Ove Røkholt, organisationsforskare: 'Dagens ko-

operation schizofren.''' In *Vår idé och framtiden. See* LRF 1986a.
—. 1987. "Kooperativ särprägel. Grunden för kooperativ förnyelse." *Kooperativ årsbok:* 10—25.
Rokkan, Stein. 1966. "Norway: Numerical Democracy and Corporate Pluralism." In *Political Oppositions in Western Democracies. See* Dahl, ed.
Roos, Leslie L., Jr., and Frederick A. Starke. 1981. "Organizational Roles." In *Handbook of Organizational Design. See* Nystrom and Starbuck, eds.
Rothstein, Bo. 1986. *Den socialdemokratiska staten. Reformer och förvaltning inom svensk arbetsmarknads- och skolpolitik.* Lund: Arkiv.
Ruin, Olof. 1968. *Mellan samlingsregering och tvåpartisystem. Den svenska regeringsfrågan 1945—1960.* Stockholm: Bonniers.
Samuelsson, Kurt. 1987. *Att tro på människorna. Möte med Bengt Westerberg.* Stockholm: Rabén & Sjögren.
Schiller, Berndt. 1975. "Years of Crisis, 1906—1914." In *Sweden's Development from Poverty to influence. See* Koblik, ed.
Schmidt, Carl. (1919) 1986. *Political Romanticism.* Translated by Guy Oakes. Cambridge, Mass.: MIT Press.
Schonfield, Andrew. 1965. *Modern Capitalism. The Changing Balance of Public and Private Power.* London: Oxford University Press.
Self, Peter, and Herbert J. Storing. 1962. *The State and the Farmer.* London: George Allen and Unwin.
Seyler, Hans. 1983. *Hur bonden blev löhearbetare. Industrisamhället och den svenska bondeklassens omvandling.* Lund: Arkiv.
Simonson, Birger. 1985. *Socialdemokratin och maktövertagandet. SAP:s politiska strategi 1889—1911.* Gothenburg: University of Gothenburg, Department of History, Meddelanden från historiska institutionen, No. 28.
Sköld, Lars, and Arne Halvarson. 1966. "Riksdagens sociala sammansättning under 100 år." *Samhälle och Riksdag.* Vol. 1. Stockholm: Almqvist & Wiksell.
Smith, Martin J. 1988. *Consumers and British Agricultural Policy: A Case of Long-Term Exclusion.* Essex: Essex Papers in Politics and Government, No. 48.
Söderlind, Donald, and Olof Petersson. 1986. *Svensk förvaltningspolitik.* Uppsala: Diskurs.
Steen, Anton. 1980. "Bønders klassintresser og offentlig politikk." *Tidsskrift for samfundsforskning* 21: 395—416.
—. 1981. "The Farmers, the Consumers and the State: Redistribution and Conflicts in Norway, Sweden and the United Kingdom." *European Journal of Political Research* 9: 1—16.
—. 1988. *Landbruket, staten og sosialdemokratene. En komparativ studie av interessekonflikter i landsbrukspolitikken i Norge, Sverige og England 1945—1985.* Olso: Universitetsforlaget.
Stenkvist, Jan. 1987. "Sven Hedin och Bondetåget." In *Heroer på offentlighetens scen,* by Knut Johanssen, et al. Stockholm: Tiden.
Stjernqvist, Nils. 1975. "The Creation of the 1809 Constitution." In *Sweden's Development from Poverty to Affluence. See* Koblik, ed.
Stjernqvist, Nils, ed. 1989. *The Swedish Riksdag in an International Perspective.* Stockholm: Bank of Sweden Tercentenary Foundation.
Stolpe, Herman. 1976. *Konsument- och producentkooperation. Tillhör de samma rörelse?* Stockholm: Rabén & Sjögren.
Strandberg, Fredrik. 1968. "Bondetåget 1914." *Från storstrejken 1909 till folkhems-*

politik. Den svenska historin 14: 70—72. Stockholm: Albert Bonniersförlag.
Stråth, Bo. 1987. *The Politics of De-Industrialization. The Contraction of the West European Shipbuilding Industry*. London: Croom Helm.
Sundsbo, Stein. 1971. "Jordbrukets faglige organisasjoner og samfunnsstrukturen i de norske bygdene." *Senit. Senterbevegelsens ideologiske tidsskrift* 2: 14—19.
Svensson, Claes. 1985. "Enhetlig eller federativ organisation —strategiskt val för lantbrukskooperationen." In *Landbrugskooperationen i Norden. See* Bager, ed.
Swedborg, Erik. 1980. *Lantbrukspolitik för 80-talet. Besluten 1977—79. Bakgrund, riktlinjer och tillämpning*. Stockholm: LTs förlag.
Thullberg, Per. 1972. "Bönderna blir konsumenter. RLF och kohandeln." *Ord och Bild* 3: 144—49.
—. 1974. "SAP och jordbruksnäringen 1920—1940. Från klasskamp till folkhem." *Arbetarrörelsens årsbok*. Stockholm: Prisma.
—. 1977. *Bönder går samman. En studie i Riksförbundets Landsbygdens Folk under världkrisen 1929—1933*. Stockholm: LTs förlag.
—. 1980. "Swedish Agricultural Policy During the Postwar Period. A Study in Rationality." Paper for the European Consortium of Political Research Joint Sessions, Florence, Italy, 25—30 March.
—. 1988. *Solidariteten sätt på prov. Mjölkskrejken i Dalarna 1942*. Stockholm: University of Stockholm, Department of History.
Thygesen, Joh:s. 1925. "Jordbruksdepartementets förhistoria och tillkomst." In *Kungliga jordbruksdepartementet 1900—1925. See* Jo.
Tiderius, Gert. 1977. *Effekten av den statliga rationaliseringspolitiken. En studie av lantbruksnämndernas medverkan i jordbrukets strukturrationalisering 1950—1970*. Lund: Studentlitteratur.
Tingsten, Herbert. 1941. *Socialdemokratins idéutveckling* I. Stockholm: Tidens förlag.
Torpe, Tjia. 1982. *Agrara opinioner i samband med 1960s års jordbruksutredning*. Stockholm: University of Stockholm, Department of History. Photocopy.
Torstendahl, Rolf. 1969. *Mellan nykonservatism och liberalism. Idébrytningar inom högern och bondepartierna 1918—1934*. Uppsala: Almqvist & Wiksell.
Uddman, Paula. 1986. Unpublished statistics. Lund: University of Lund Department of Political Science.
Urwin, Derek W. 1980. *From Ploughshare to Ballotbox. The Politics of Agrarian Defence in Europe*. Oslo: Universitetsforlaget.
Utterström, Carl. 1980. *Organizational Visions, Ideologies and the Cooperative Myth. A Study of Mergers Among Farmer Cooperatives in Sweden*. Uppsala: University of Agricultural Science, Department of Economics and Statistics.
—. 1982. "Medlemsdemokrati i stora lantbrukskooperativa föreningar." In *Landbrugskooperationen i Norden. See* Als and Møgelhøj, eds.
Verba, Sidney, et al. 1987. *Elites and the Idea of Equality. A Comparison of Japan, Sweden and the United States*. Cambridge, Mass.: Harvard University Press.
Viklund, Daniel. 1987. *Att förstå EG. Sveriges framtid i Europa*. Stockholm: Norstedts.
Wentzel, Ann-Kristin, et al. 1980. *Hur sårbart är Sverige*. Stockholm: Sekretariatet för Framtidsstudier.
Wilensky, Harald L. 1967. *Organizational Intelligence. Knowledge and Policy in Government and Industry*. New York: Basic Books.
Wilson, Graham K. 1977. *Special Interests and Policymaking. Agricultural Policies*

in Britain and the United States of America, 1956—70. London: John Wiley.

Wilson, James Q. 1973. *Political Organizations*. New York: Basic Books.

Winters, L. Alan. 1986. *Industrial Countries' Agricultural Policy: How, What and Why*. London: Centre for Economic Policy Research, Discussion Paper Series, No. 118.

Wohlin, Nils. 1919. *Faran af bondeklassens undergräfvande. I sammanhang med de gamla arfvejordsåskådningarna upplösning, emigrationen och bondejordens mobilisering*. Stockholm: Nordstedt.

"The Political Impasse in Farm Support Legislation." 1962. *Yale Law Journal* 71: 952—78.

Young, Birgitta. 1987. "The American Dairy Industry: Meso-Corporatism and its Contradictions." Paper for the European Consortium of Political Research, Amsterdam, Holland, 10—15 April.

Zetterberg, Hans L. 1978. "Notes on Environmental Awareness and Political Change in Sweden." Paper for the conference "Environmental Awareness and Political Change," Berlin, 9—10 January.

Index

About the Author

Michele Micheletti is assistant professor of political science at the University of Stockholm, Sweden. She was born in the United States but has resided in Sweden for the last 15 years. She received her B.A. from the University of California at Santa Barbara and her Ph.D from the University of Stockholm. She has written on Swedish politics and interest organizations. The peak labor market organizations have been her main concentration. The research themes have been member democracy and the close cooperation between the state and interest organizations. Presently she is researching a wide range of interest organizations, ad hoc groups, and popular movements in Sweden.

www.ingramcontent.com/pod-product-compliance
Ingram Content Group UK Ltd.
Pitfield, Milton Keynes, MK11 3LW, UK
UKHW020812290926
15470UKWH00003B/1100

9 780275 933982